COLOUR DICTIONARY OF

TREES & SHRUBS

FRASER STEWART

This edition published in 1994 by
Fraser Stewart Book Wholesale Ltd
Abbey Chambers
4 Highbridge Street
Waltham Abbey
Essex EN9 1DQ

Produced by Marshall Cavendish
Books, London

Editor: Susanne Mitchell
Art editor: Caroline Dewing
Designer: Sheila Volpe

ISBN 1 85435 725 5

British Library Cataloguing in
Publication Data:
A catalogue record for this book is
available from the British Library

Printed and bound in Malaysia

Introduction

This dictionary is more than a reference book of names, heights, and dates. It contains practical cultural advice to help you get the best from your plants. And if you are uncertain of a particular method of propagation, want to know about the pros and cons of different composts, or are not sure about a particular term or expression used, simply refer to the second part of the book where the section on terms and techniques should help you.

It has not been possible, in a book of this size, to include all the species, let alone varieties, that you might find in catalogues and garden centres. In some cases there are hundreds of species that we have had to compress into half a dozen. But the same cultural advice generally applies to most of them, and we have included as examples those species that you are most likely to find. The chances are, unless you buy an unusual plant, you will find it within these pages. And even if the exact species is not listed, the general advice is still likely to apply.

To be able to cover shrubs and broad-leaved trees more comprehensively we have had to exclude conifers from this book. And some of the larger, parkland-type trees have also been excluded in favour of those more suitable for gardens.

The aim has been to take a practical approach, from the plants included to the cultural advice. We have not always included plants under their latest botanical names (although they are given in the text), but have listed them under the names most often used by gardeners. It takes many years—up to a generation—for a change in botanical nomenclature to work its way through to the commercial trade, and some of the trees and shrubs in this book are rarely sold under their 'correct' name. With old synonyms, too, we have listed only those under which a particular plant might still occasionally be listed.

Flowering times and activities have been given as seasons, as the months will vary from country to country. As a guide, the seasons have been taken to mean the following in the northern and southern hemispheres respectively:

Early spring: March/September
Mid spring: April/October
Late spring: May/November
Early summer: June/December
Mid summer: July/January
Late summer: August/February
Early autumn: September/March
Mid autumn: October/April
Late autumn: November/May
Early winter: December/June
Mid winter: January/July
Late winter: February/August

What the symbols mean

It is often tempting to buy a plant because you like the look of a picture, or on impulse in the garden centre or nursery, but you should at least know whether it has much chance of success. Two important considerations are whether the soil is very acid or alkaline, and

Mixed shrub border with roses and lavender

One of the attractive deutzia hybrids

whether you can provide a position in sun or shade (or possibly partial shade). Most plants will tolerate a wide range of soils, and only relatively few will react badly to strongly acid or to alkaline soils—but these include some important plants, such as rhododendrons and lilacs. The soil and aspect required are given in symbols at the beginning of each entry, so you can easily see whether the plant is likely to be suitable.

The following at-a-glance symbols have been used throughout this book:

TR = tree
SH = shrub
Cl = climber
(E) = evergreen
FS = best in full sun
PS = best in partial shade (dappled light or full sun for part of day)
SD = will tolerate shade (for those with this symbol alone, requires a shady position)
AD = requires an acid soil (low pH)
AK = does best on an alkaline soil (high pH)
O = will tolerate a wide range of soils
M = best in moist position
D = thrives in dry soils

Heights

Heights have not been given as precise measurements. It is impossible to predict tree and shrub heights accurately for all soils and gardens. Height will depend on where you live (the overall climate), the soil, aspect, and proximity to other trees and shrubs. Even how you prune and train them. Both rate of growth and ultimate height can vary enormously. However, some comparative guide is both desirable and necessary. We have therefore given likely height ranges after 15 years. In the case of most shrubs this will also be the ultimate height, but some trees may of course go on to grow larger. However, 15 years is as far ahead as most people plan their garden, and after this time growth is likely to slow down considerably.

The following descriptions have been used throughout the book.

Shrubs
Prostrate: ground-hugging to 30 cm (1 ft)
Dwarf: 30–60 cm (1–2 ft)
Small: 60 cm–1.5 m (2–5 ft)
Medium: 1.5–3 m (5–10 ft)
Large: 3 m+ (10 ft+)

Trees
Very small: up to 3 m (10 ft)
Small: 3–4.5 m (10–15 ft)
Medium: 4.5–7.5 m (15–25 ft)
Large: 7.5–15 m (25–50 ft)
Very large: 15 m+ (50 ft+)

The Plants

Abelia

SH(E, some)/O/FS

A genus of slightly tender evergreens and deciduous shrubs named to commemorate Dr Clarke Abel (1780–1826), a physician attached to Lord Amherst's mission to China in 1816–17.

How to grow
Abelias revel in full sun, and also do best near a warm, sunny wall in all but the most favourable areas. They are not dependable in cold districts unless you can provide a sheltered position, but if you can provide a suitable spot they are worth trying because of their long flowering period.

Abelia x *grandiflora*

Propagation
Cuttings of semi-ripe shoots, preferably with a heel, taken in mid or late summer. Shade from direct sun, and root under glass at about 18–21°C (65–70°F). They should root in a month, then pot up singly and grow on at 13–18°C (55–65°F) for another month. Overwinter in a frost-free frame or cool greenhouse. Pot on into 13 cm (5 in) pots in spring. Harden off and move to a sheltered position outdoors when risk of frost is past. Plant out in the chosen position in the autumn.

SOME POPULAR SPECIES	
A. chinensis *(China)* Spreading deciduous shrub with fragrant, white, tubular flowers flushed rose, usually borne in pairs, from mid summer to early autumn. Only hardy in mild areas. Small. **A. floribunda** *(Mexico)* Semi-evergreen producing an abundance of cherry-red tubular flowers in early summer. Only suitable for mild areas. Small. **A. x grandiflora** *(Garden origin)* A wide-spreading semi-evergreen with slender, arching stems clothed with small dark green leaves. Slightly fragrant, funnel-shaped white flowers tinged pink, from mid summer to early autumn.	Hardy in mild areas, but may be cut back to ground level in a severe winter. Small to medium. **A. schumannii** *(China)* A bushy plant with dark green leaves. Long, lilac-pink flowers about 2.5 cm (1 in) long from early summer to early autumn. Not dependably hardy even in mild districts. Small to medium. **A. triflora** *(Himalayas)* Upright, bushy shrub with lance-shaped dull, dark green leaves. White, flushed rose, fragrant flowers borne in erect clusters 5 cm (2 in) wide, usually with three flowers on each stalk, in early summer. Perhaps the hardiest abelia. Medium to large.

Acer

Maple

TR–SH/O, M/FS–PS

There are over 100 species. Some are forest trees with no place in the average garden, but many of the Japanese maples are slow-growing and compact. A few have brilliant leaves in spring, others assume vivid autumn colours, and there are forms with purple leaves all summer. Some have interesting barks that make the tree attractive even in winter.

How to grow

Acers prefer a moist but not wet soil, and do less well on alkaline soil though they can be grown.

The tree forms will obviously tolerate and enjoy full sun, but the Japanese maples *(A. palmatum)* and *A. japonicum* are generally best in partial shade. Some yellow-leaved varieties will scorch in full sun.

Species grown for their decorative bark are best positioned where winter sunlight will fall on them.

Be sure to cut out any green-leaved branches that appear on variegated varieties.

Propagation

Can be propagated from seed sown in a cold frame in mid autumn (soak in tepid water for 48 hours first). Germination may take up to 18 months. Pot up the seedlings singly, and overwinter in a frame for the first couple of years. Pot on in the second spring and plant out in the third year.

Named varieties and coloured and variegated forms must be grafted. This is not easy for an amateur.

Acer pseudoplatanus 'Brilliantissimum'

SOME POPULAR SPECIES

A. campestre *(Europe, Western Asia)*
(Field maple) A round-headed tree, usually low-branching. Five-lobed leaves. Long-lasting autumn colour. Tolerates chalk well, and even wet soil. Medium tree.

A. capillipes *(Japan)*
(Snakebark maple) Upright tree with three-lobed mid-green leaves, reddish when young, turning crimson in autumn. Trunk and branches with longitudinal white stripes. Will form a domed, spreading head. Large tree.

A. griseum *(China)*
(Paperbark maple) Beautiful and distinctive tree, famed for its buff or brown bark that peels to reveal cinnamon-coloured underbark. The mid green leaves with three leaflets assume fiery tints in autumn. Medium tree.

A. japonicum *(Japan)*
Slow-growing small tree or large, rounded bush. Two good varieties are 'Aureum' (soft yellow, lobed leaves, retaining colour all summer) and 'Aconitifolium' (deeply lobed leaves with brilliant autumn reds). Small tree or large shrub.

A. negundo *(North America)*
(Box elder) Wide-spreading head on a low-branching tree. Long-stalked mid green leaves with three to five leaflets. Choice varieties are 'Auratum' (golden leaves throughout summer). 'Elegans', syn 'Elegantissimum' (bright yellow margins), and 'Variegatum' (irregular white margins). Medium to large tree.

A. palmatum *(Japan)*
(Japanese maple) Round-headed large shrub or small tree with lobed bright green leaves that turn brilliant red and orange in autumn. Widely-grown varieties include 'Dissectum' (deeply-cut green leaves, red in autumn), and 'Dissectum Atropurpureum' (bronze-purple deeply-cut leaves, red in autumn). Small tree, very useful for the smaller garden.

A. pensylvanicum *(North America)*
(Snakebark maple) Erect tree with three-lobed mid green leaves turning yellow in autumn. Grown mainly for its bark, which is striped with jagged white lines (these may not be seen on young trees). Medium tree.

A. platanoides *(Europe)*
(Norway maple) Fast-growing large tree with five-lobed bright green leaves turning yellow in autumn. Varieties include 'Crimson King' (deep crimson-purple), and 'Drummondii' (white margins). Large or very large tree.

A. pseudoplatanus *(Europe)*
(Sycamore) Five-lobed green leaves, turning yellow in autumn. More desirable varieties include: 'Brilliantissimum' (slow-growing, leaves unfolding in shades of pink, later turning green), 'Leopoldii' (variegated yellowish-pink speckled white or cream), and 'Worlei' (the golden sycamore, foliage yellow becoming green later in season). Very large tree (some of the varieties smaller).

Actinidia

CL/O/FS

A genus of 40 deciduous twining climbers. The species described below are grown for quite different reasons. *A chinensis* is grown mainly for its fruit (kiwi fruits), though you should not expect too much from these outdoors. *A. kolomikta* is grown for its strikingly variegated leaves.

How to grow

If you want to grow kiwi fruit, you will need to plant male and female plants in close proximity. These are more successful in a protected environment. If growing merely as a climber bear in mind that it will be vigorous and needs plenty of space.

Actinidias will not do well on chalky or badly drained ground.

A. chinensis will need a high wall, with a suitable support, or a tree to grow into. Pinch out the growing tip when young to encourage bushy growth. *A. kolomikta* has a more refined habit and will grow best against a sheltered wall or fence. Full sun will bring out the best of the unusual leaf variegation.

Propagation

A. chinensis can be grown from seed sown in a cold frame in mid autumn. Prick the seedlings out singly in spring, using 10 cm (4 in) pots, and grow on in a garden frame. Harden off and plunge the pots in a sheltered spot outdoors in early summer. Plant out in autumn.

Both species can be raised from cuttings (this method will enable you to know whether you are propagating male or female plants of *A. chinensis*). Take semi-ripe cuttings in mid summer, and root in warmth, preferably in a mist propagator. Pot up into 13 cm (5 in) pots when rooted (it will take about a month), and place in a cold frame to overwinter. Harden off in late spring and plunge outside from mid summer, ready to plant in the autumn.

Suitable branches can be layered in the autumn and severed a year later if rooted (otherwise leave for another year). This is an easy method for a couple of plants.

Actinidia kolomikta

SOME POPULAR SPECIES	
A. chinensis *(China)* (Kiwi fruit, Chinese gooseberry) Vigorous twining climber with heart-shaped dark green leaves. Creamy-white, cup-shaped flowers 4 cm (1½ in) across in summer, followed by edible egg-shaped fruits covered with reddish-brown hairs, if male and female plants are present. If grown for fruit, it is best to obtain a clone	selected for its fruit. **A. kolomikta** *(Manchuria, China, Japan)* (Kolomikta vine) Slender climber with oval to oblong or heart-shaped dark green leaves with pink or white tips—the main attraction of the plant, a feature at its best in full sun. Slightly fragrant white flowers in early summer. Moderately vigorous climber.

Aesculus

TR–SH/O/FS–PS

A genus of 13 species of trees and shrubs, including the well-known horse chestnut (conker tree) of Britain, and the American buckeyes. They are very ornamental trees, flowering in late spring and early summer.

Some will be far too large for most gardens, but some of those listed below are of much more modest proportions.

How to grow

Provided you give the species sufficient space, they will look after themselves. They do less well on chalky soil.

Propagation

To raise from seed, prepare the 'conkers' by placing in trays of sand, covered by about 2.5 cm (1 in), and leave them in a sheltered place outdoors for the winter. In spring set the conkers about 5 cm (2 in) deep and 15 cm (6 in) apart in a nursery bed. Let the seedlings grow for two or three years, then plant in the autumn.

Hybrids and named varieties that do not come true from seed must be grafted, usually using three-year-old *A. hippocastanum* as the rootstock, but this is really a job for the specialist.

SOME POPULAR SPECIES
A. x carnea *(Garden origin)* (Red horse chestnut) Round-headed tree. Large, mid green leaves with five to seven leaflets. Reddish-pink 15–20 cm (6–8 in) 'candles' in late spring or early summer. The colour is deeper in 'Briotii'. Large tree. **A. hippocastanum** *(Balkans)* (Horse chestnut) The well-known 'conker' tree. Wide-spreading head. Large, mid to dark green palmate leaves with five to seven leaflets. 'Candles' up to 30 cm (1 ft) long, the white flowers blotched red, in late spring. Very large tree. **A. indica** *(Northern India)* (Indian horse chestnut) Short-trunked tree with 'candles' of white flowers fringed pale pink and blotched yellow and red, in early or mid summer. The shiny conkers are black. Large to very large tree. **A. parviflora** *(USA)* (Dwarf buckeye) A broad, bushy shrub that spreads by suckers. White 'candles' with pink stamens in mid summer. Medium to large shrub. **A. pavia** *(North America)* (Red buckeye) A round-headed shrub or small tree. 15 cm (6 in) panicles of crimson flowers. Small tree or large shrub.

Aesculus x *carnea* 'Briotii' *is a more compact form of the red horse chestnut*

Ailanthus

Tree of Heaven

TR/O/FS

The name of this genus comes from a native word for a tropical species (*A. moluccana*): *ailanto*, which means 'a tree tall enough to reach the skies'. The common name for the most widely grown species also reflects its upright, fast and tall-growing nature.

How to grow

A very easy tree that will tolerate atmospheric pollution well, and grow in any kind of soil. No special treatment is required. But avoid digging around the roots as this encourages suckers.

Propagation

Seed can be sown in pots in a cold frame in mid autumn. Prick out seedlings into a nursery bed or into pots in spring or early summer. Plant in the autumn.

Root cuttings should be taken in late winter. Dig up pencil-thick roots and cut into sections about 5 cm (2 in) long. Insert vertically in pots and root at about 21°C (70°F). Pot up singly when rooted and the tops have made growth. Grow under protection for a few weeks more then harden off and stand outdoors in a shaded spot for the summer. Plant in autumn.

Ailanthus altissima

SOME POPULAR SPECIES	
A. altissima *(China)* (Tree of Heaven) Fast-growing, round-headed tree with ash-like leaves	up to 1 m (3 ft) long with 15–30 leaflets. Reddish-brown 'keys' appear on female trees. Very large.

Akebia

CL/O/FS–SD

A genus of five vigorous semi-evergreen twining climbers. Suitable for training over porches or on a trellis.

How to grow

Best on a light soil. Although hardy, need a mild spring and a long hot summer to set their unusual, sausage-shaped fruits. Evergreen in mild, warm areas, but otherwise semi-evergreen.

Cut back hard every three or four years.

Akebia quinata

Propagation

Sow seed in warmth under glass in late winter (germination may take up to six months). Prick out into pots and harden off. Overwinter in a cold frame, pot on in spring, and plant out in the autumn. Semi-ripe cuttings taken in mid summer and rooted in warmth should be successful. Pot up singly then treat as for seedlings.

An easy method for just a few plants is to peg down layers in the autumn. They will be ready to sever the next autumn, or the year after.

<table>
<tr><td colspan="2">SOME POPULAR SPECIES</td></tr>
<tr><td>A. quinata (Japan, China, Korea)
A twining climber that can be grown as a spreading or trailing shrub. Glabrous leaves are formed of five long-stalked leaflets. Fragrant, pendent purple flowers appear in mid</td><td>spring, separate male and female flowers are carried on the same raceme. May be followed by violet fruits. A fairly vigorous climber.
A. trifoliata (Japan, China)
Similar to above but with trifoliate leaves.</td></tr>
</table>

Alnus

TR/O, M/PS–FS

A genus of about 30 trees and shrubs, with male and female catkins on the same tree. Most of them flower in spring. The majority are fast-growing and useful for damp and heavy soil.

How to grow

These tough plants will grow happily with the minimum of attention. They do not usually do well on shallow chalky soil.

Propagation

Sow seed in containers in a cold frame in mid winter. After a month bring them into a temperature of 16–18°C (60–65°F) for another month, by which time they should have germinated. Pot up the seedlings and overwinter in a cold frame and pot on in spring. Harden off, stand outdoors, and plant in the autumn.

An easier method is to remove suckers in late autumn and plant where they are to grow.

Cuttings of ripe wood taken in late autumn should root in a cold frame and be ready to plant out a year or two later.

Alnus cordata

<table>
<tr><td colspan="4">SOME POPULAR SPECIES</td></tr>
<tr><td>A. cordata (Corsica, Southern Italy)
(Italian alder) Pyramidal tree with roundish or broadly oval shiny green leaves. The fruiting heads are erect and cone-shaped. Fast-growing and</td><td>one of the most lime-tolerant species. Large to very large.
A. glutinosa (Europe, Western Asia, North Africa)
(Common alder) Narrow, pyramidal tree with smooth grey bark and</td><td>pear-shaped serrated, shiny green leaves. Yellow catkins in early spring. Very large. Several good varieties.
A. incana (Europe, Caucasus)
(Grey alder, white alder,</td><td>hoary alder) Wide-trunked and spreading tree with oval to round leaves, dull green on top, grey beneath. Fast-growing and very hardy: ideal for a cold, wet position. Very large.</td></tr>
</table>

Amelanchier

Snowy mespilus, June berry, shad bush

TR–SH/O/PS–FS

Amelanchier lamarckii

A genus of about 35 very hardy deciduous shrubs or trees, most of which combine the merit of attractive flowers in spring with autumn colour.

The most common species are described here, and there is much confusion between them. The plant sold as *A. lamarckii* may be *A. canadensis* and vice versa, and to add to the complication, some botanists consider that the plant sold in some countries (including Britain) as *A. canadensis* is in fact often *A. laevis*. *A. x grandiflora* belongs with *A. lamarckii*.

Let none of this worry you—they are all very similar and equally desirable.

You may also be confused to see the plant sometimes described as a tree, sometimes as a shrub. The choice is yours: it can be grown as a trunked tree or allowed to branch when young and grown as a large shrub.

How to grow

Although amelanchiers will grow on most soils, they are likely to do less well on chalk.

They need no regular pruning or other attention, but you need to decide whether you want a tree or shrub when you buy as the initial training will already have been done.

Propagation

Stratify or prechill seed before sowing in a cold frame in spring. Overwinter under a frame, then harden off and move to a nursery bed. Grow on for a couple of years before planting out.

An easier method is to lift suckers from an established plant. Large ones can be planted immediately into their permanent position, small ones are best grown in a nursery bed for a year.

Layers provide another quick and easy means of propagation. Peg them down in early spring or early autumn, and lift a year later.

SOME POPULAR SPECIES	
A. canadensis *(North America)* Masses of white star-shaped flowers in erect loose heads in mid spring. May be followed by sweet-tasting, edible berries in early summer. Small, rounded or oblong mid	green leaves colouring red or yellow in autumn. Large shrub or medium tree. **A. lamarckii** *(North America)* Similar to above species. The differences are minor.

Aralia

Japanese angelica tree

TR–SH/O/PS–FS

A genus of hardy or nearly hardy trees, shrubs and even herbaceous plants.

How to grow

Will do best on light, well-drained soil. Can be grown as a single-stemmed tree or as a multi-stemmed plant when it will be more shrub-like. No special care needed, but if shoots die back, prune hard in spring to stimulate growth.

Aralia elata

Propagation

Seed can be sown in warmth in early spring (germination usually takes one to three months). Prick out into pots, then gradually harden off in a cold frame, where the seedlings can spend the first winter. Grow on in a nursery bed for another year.

Suckers can be difficult to establish. Pot them up in late autumn or early spring, and grow them on in a frame. Never let them dry out. Harden them off and stand outside for the summer. Plant in autumn.

Root cuttings are not difficult and can be taken in late winter (see Ailanthus).

SOME POPULAR SPECIES	
A. elata *(North East Asia)* Large bi-pinnate leaves sometimes 90 cm (3 ft) long. Small, whitish flowers appear in large heads 30–60 cm (1–2 ft) across in late summer or	early autumn. There are variegated varieties: 'Aureovariegata' has yellow margins, becoming silvery-white with age; 'Variegata' has leaves edged white.

Arbutus

Strawberry tree

TR – SH(E)/O/FS

A genus of 20 hardy and half-hardy evergreen trees and large shrubs. The common name stems from the fruits that resemble strawberries in appearance. These are unlikely to be plentiful and may take about a year to ripen to orange-red, so you could have flowers and coloured fruit at about the same time. Although they belong to the Ericaceae family they are lime tolerant, especially *A. unedo* and *A.* x *andrachnoides.*

Arbutus x *andrachnoides*

How to grow

Most species are best on lime-free soil. Will need shelter from cold north and east winds. Young plants will need protection until they are well established.

Propagation

Seed needs prechilling for a month, before surface sowing in late winter. Give them the protection of a frame or cold greenhouse. Germination may take up to six months. Pot up into an ericaceous compost, and grow in a frame for a couple of years before planting out.

Cuttings are fairly easy to root if you use semi-ripe wood with a heel, taken in mid summer. Use a lime-free compost. They should root in about a month, and can then be treated like seedlings.

SOME POPULAR SPECIES	
A. andrachne *(South East Europe)* (Grecian strawberry tree) Tender when young, Grown chiefly for its smooth, cinnamon-coloured bark. Small white flowers in spring. Orange-red fruits follow. Large shrub or small tree. **A. x andrachnoides** *(A. x hybrida) (Garden origin)* Distinctive peeling cinnamon-red bark. Clusters of ivory-white flowers in winter or spring. Fruit seldom formed. Medium tree. **A. menziesii** *(California to British Columbia)*	(Madrona) Smooth, terracotta-coloured bark. Pitcher-shaped white flowers in mid or late spring. Orange-yellow fruits. Hardy only in favourable areas. Medium to large tree. **A. unedo** *(Asia Minor, Southern Europe, South West Ireland)* (Killarney strawberry tree) Wide-topped tree or large shrub, with gnarled trunk displaying rough, shredding brown bark. Pitcher-shaped white flowers mid autumn to early winter. Orange-red fruits. Medium tree.

Aucuba

Spotted laurel
SH(E)/O/FS–SD

A genus of hardy evergreen, shade-loving shrubs. Female plants produce large red berries in winter if a male plant is nearby. Some varieties are only male, others only female (cross-pollination with other varieties should ensure the production of berries).

Aucuba japonica 'Variegata'

How to grow
One of the most successful shrubs for a difficult site—tolerating dense shade and industrial pollution. Although shade-tolerant, it can be grown in sun (better for variegated varieties).

Pruning is not necessary but the shrub will stand cutting back severely if necessary.

Propagation
Seed can be sown in a cold frame in autumn, or stratified then sown late winter. Grow seedlings on in pots in the frame for a couple of years (shade in summer) then plant out in autumn or spring.

Cuttings are quicker, and enable you to be sure of getting male or female plants. Take 10–15 cm (4–6 in) semi-ripe heel cuttings in early autumn and root in a cold frame. Pot up in spring if rooted and plant out the following spring.

SOME POPULAR SPECIES
A. japonica *(Japan)* Rounded, bushy shrub with large, leathery, oval, green leaves. Variegated varieties, usually splashed or blotched yellow, of which there are many, are more attractive. Insignificant olive-green flowers in spring; clusters of oval, bright scarlet berries on female plants from autumn to spring. Medium.

Berberis (evergreen)

Barberry
SH(E)/O/PS–FS

An invaluable group of shrubs, with hundreds of species (some of the deciduous species are described in the next entry). Most species have orange or yellow flowers in spring, often followed by blue or blackish berries in autumn.

How to grow
Berberis thrive on most soils, but do best in an open, sunny position. The evergreen species tolerate shade better than the deciduous kinds. They form neat, usually round, bushes and need little pruning. Just thin out overcrowded shoots after flowering to retain the desired shape and remove any dead wood.

Berberis darwinii

Propagation

Seed can be sown in late winter after stratification or prechilling. Given about 16–18°C (60–65°F) they should germinate in one to two months. Grow on in pots in a frame for a year, then plunge outdoors for another year before planting. *Seed-raised plants are often variable.*

Taking cuttings is a more popular method. Take semi-ripe heel cuttings in autumn and root in a cold frame. Pot up and harden off before plunging outdoors for a couple of years before planting.

Layers are another possibility—see deciduous species.

SOME POPULAR SPECIES	
B. darwinii *(Chile)* (Darwin's barberry) Dense, bushy shrub. Drooping clusters of small orange-gold flowers in mid and late spring. Medium to large. **B. gagnepainii** *(China)* Erect, dense shrub. Clusters of yellow flowers in late spring, followed by red berries. Medium. **B. linearifolia** *(Chile)*	Erect, slow-growing and loose shrub. Rich orange flowers in mid spring, followed by black, oval berries. May be vulnerable in cold districts. Medium. **B. x stenophylla** *(Garden origin)* Dense, thicket-forming shrub with long, arching branches. Purple berries. Makes an impenetrable hedge. Medium.

Berberis (deciduous)

Barberry

SH/O/FS

Although the evergreen berberis (see above) provide year-round interest, the deciduous species are among the most spectacular, many having vivid autumn colour.

How to grow

Will thrive on most soils, but prefer an open, sunny situation. Prune in late winter, just shortening straggly shoots to shape the bush. Cut out old, dead, or diseased stems.

Berberis thunbergii 'Atropurpurea Nana'

Propagation

Seeds and cuttings can both be used—see evergreen species.

An easy method is to layer shoots in early or mid spring. Two-year-old shoots should root readily if kept moist, and be ready to sever in about two years. Lift with a big root-ball and plant out straight into the permanent position.

Division is more 'instant'. Lift and split up young plants in autumn or spring. Alternatively remove well-rooted vigorous suckers from the outside edges of old, long-established clumps, and replant immediately. Keep moist until established.

SOME POPULAR SPECIES	
B. x ottawensis *(Garden origin)* Oval or rounded, leathery, glossy green leaves and pendulous clusters of red berries. 'Superba', with upright habit and arching branches carrying deep purple, oval leaves, is perhaps the best form. Small to medium. **B. x rubrostilla** *(Garden origin)* Compact shrub with graceful arching stems. Leaves become brilliantly coloured in autumn. Yellow flowers in late spring, followed by large, showy, coral-red berries. Small to medium.	**B. thunbergii** *(Japan)* (Thunberg's barberry) Pale yellow flowers, usually suffused red, in mid and late spring. Bright red berries in autumn. Brilliant autumn foliage. 'Atropurpurea' and 'Atropurpurea Nana' are popular varieties. The species makes a small or medium shrub, some varieties are dwarf. **B. wilsoniae** *(China)* (Mrs Wilson's barberry) Yellow flowers in mid summer, followed by coral red berries. A spreading habit, excellent for clothing a bank. Dwarf to small.

Betula

Birch

TR/O/PS–FS

The birches naturally inhabit some of the most inclement regions of the world and are therefore very hardy trees. They are all deciduous. The popular *B. pendula* will thrive even on very poor soils and will be among the first trees to re-colonise an area left to go wild. Despite this, there are some beautiful birches, many with very attractive bark. Catkins are a bonus in spring, although these are not a particular feature.

How to grow

Although birches are very tough, they are shallow-rooting, so avoid digging beneath them, and in times of drought they may need water-ing—especially if young. Otherwise they need no attention.

Propagation

Seed can be sown thinly in containers in a cold frame in spring (just press the seeds gently into the surface of the compost). Shade from strong sun. The following spring, pot up the seedlings or plant in a nursery bed. Grow on for a couple of years before planting out.

Grafting is used by professionals for hybrids and varieties that do not come true from seed. Saddle graft in spring, using three-year-old stems in autumn. They should be ready to sever in about two years.

SOME POPULAR SPECIES

B. ermanii *(China, Korea, Japan)*
Peeling creamy-white bark tinted pink. Branches assume shades of orange-brown. Broadly ovate, mid green leaves with heart-shaped base. Best in mild areas because it starts into growth early. Large.

B. jacquemontii *(Himalayas)*
Similar to other birches but grown for its white, peeling bark. Large.

B. papyrifera *(North America)*
(Canoe birch, paper birch) Another birch grown mainly for its bark, which is white and peels away in large strips on well-established trees. Large.

B. pendula *(B. alba, B. verrucosa) (Europe)*
(Silver birch) Silvery-white trunk. Pendulous branches with small, somewhat diamond-shaped leaves. Especially good are 'Dalecarlica' (Swedish cut-leaved birch; tall with some branches hanging to ground level) and 'Youngii' (Young's weeping birch; dome-shaped and weeping, ideal for a small garden). The species makes a large or very large tree.

B. utilis *(Himalayas)*
(Himalayan birch) Orange to brown or deep coppery-brown peeling bark. Large to very large.

Above *Birch grown as a specimen tree*
Right *Betula pendula* 'Youngii'

Buddleia

SH/O/FS

Botanists would have us spell this genus, named after an English botanist the Rev. Adam Buddle, buddleja. But you will almost always find it sold with the traditional spelling. The muddle probably stems from the fact that a tailed 'i' was frequently used at about the time the plant was named, so if one plays the naming game according to the rules of what was first published one can see how the incongruous 'j' appeared.

How to grow
Buddleias will grow in any reasonable soil provided drainage is adequate, and *B. davidii* in particular will thrive on chalky soil. They do best in full sun.

All three species listed here need different pruning. For *B. alternifolia,* simply prune out old shoots after flowering. *B. davidii* flowers on wood produced in the current season, so cut back to within two or three buds of the old wood in early spring. *B. globosa* flowers on both old and young wood, so prune after flowering, removing a couple of old stems and cutting back younger flowering shoots to fresh growth on the main branches.

Propagation
Buddleias from seed can be very variable and are often inferior, so it is better to use cuttings as a method of increase.

Cuttings of semi-ripe wood, 10–15 cm (4–6 in) long, should be rooted in a cold frame in mid summer. Overwinter in the frame, and pot up into 13 cm (5 in) pots in spring. Harden off and stand outdoors until the autumn, when they can be planted out.

Top *Buddleia davidii* 'Dubonnet'
Bottom *B. globosa*

SOME POPULAR SPECIES			
B. alternifolia *(China)* A tree-like shrub with lance-shaped leaves on cascading branches, which are covered with clusters of sweetly scented lavender-blue flowers in early summer. Large.	**B. davidii** *(China)* (Butterfly bush) Vigorous shrub with upright, arching shoots. The long flower heads, so loved by butterflies, appear in mid and late summer. There are many named varieties,	in shades of lilac, red-purple, violet-purple, and white. Medium. **B. globosa** *(Peru, Chile)* (Orange ball tree) Dark green, lance-shaped, wrinkled leaves about 15 cm (6 in) long. Lax	terminal clusters of fragrant, orange-yellow ball-like flowers produced in late spring or early summer. May be almost evergreen in mild areas, deciduous in cold districts. Large shrub.

Callicarpa
Beauty berry
SH/O/FS

The name of this genus is derived from two words: *kallos* (beauty) and *karpos* (fruit). The fruits are the main reason for growing these shrubs, which are otherwise not outstanding. The beauty really comes in the autumn with good leaf colour and pale purple berries.

How to grow
It is best to plant this shrub in groups rather than singly, to give more impact. They need a sheltered position to do well. As the flowers and fruits are produced on current season's growth, prune back fairly hard in late winter.

Propagation
Seed can be tried for fun, but cuttings are the normal method. Stratify or prechill before sowing in warmth in spring. Grow in a frame for two or three years before planting out.

Cuttings of semi-ripe shoots can be taken in mid summer, using a propagator. Pot up when rooted and grow in a frame for a year before planting out.

Callicarpa bodinieri giraldii

SOME POPULAR SPECIES	
C. bodinieri giraldii *(C. giraldii) (Western China)* Upright habit. Narrowly lance-shaped dull green leaves, paler beneath. Red	and yellow tints in autumn. Insignificant lilac flowers in mid summer, violet-purple berries in autumn. Medium.

Calluna
Heather, ling
SH(E)/AD/PS–FS

This is a monotypic genus—there is only one species—but it makes up for that in having hundreds of varieties. Those with golden foliage are bright and colourful the whole year. Good ground cover for an acid soil and happier in full sun although they will grow in some shade.

How to grow
Although some types of heather will tolerate some lime in the soil, the callunas must have acid conditions if they are to thrive. If necessary mulch with peat.

Cut dead heads off with shears after flowering.

Propagation
Seed can be tried for fun, though for named varieties you will need to use cuttings. Prechill for 28 days, then surface sow in spring. Germination is likely in one to two months. Grow in

trays of lime-free compost in a frame for a year. Pot up in an acid compost and grow on for another year before planting out.

Cuttings are the usual method. Take semi-ripe, 3–5 cm (1–2 in) shoots with a heel during the summer, and root in a frame or propagator. Pot up when rooted, overwinter in a frame, and grow on as seedlings.

Layers also provide a quick and easy method—see Erica.

SOME POPULAR SPECIES	
C. vulgaris *(Europe)* Stubby scale-like leaves in a wide colour range from grey and green to orange	and red. Single or double flowers in shades of purple, white, and pink. Prostrate or dwarf.

Calluna vulgaris 'Barnett Anley'

Camellia

SH(E)/A/PS–FS

George Joseph Kamel, a seventeenth century Moravian Jesuit who travelled in Asia, gave his name to some of our most beautiful flowering shrubs. They are all the more welcome because they flower early in the year.

How to grow

Avoid alkaline soil, and incorporate plenty of peat when planting. Although they will take full sun, partial shade is often better because the opening blooms may be more protected from a rapid thaw. North and east winds can be damaging, and the protection of a wall or other shrubs is usually beneficial. Despite this the plants are very tough. If planting in a container, use an ericaceous compost.

Propagation

Seed is a slow and unpredictable method of propagation. It is better to take cuttings.

Cuttings of semi-ripe wood with a heel can be taken during summer, and rooted in a propagator. They should root in about a month. Pot up singly and grow on under glass for a year.

Leaf bud cuttings can be used if you need a large number of plants.

Layers can be pegged down in early autumn, but they may take two years or more to root. Remove a sliver of bark on the undersides when you layer.

SOME POPULAR SPECIES	
C. x williamsii *(Garden origin)* Free-flowering and very hardy. Flowers February to April. There are single,	semi-double, double, anemone and paeony-flowered forms. Mainly pink shades. Plants flower young. Medium.

Camellia x *williamsii*

Caragana

Siberian pea tree

SH–TR/O/PS–FS

The name of this genus comes from *Caragan*, the Mongolian word for *C. arborescens*, the most widely planted species. Coming from Siberia, it is a tough plant able to withstand the most unpromising site, no matter how poor the soil or exposed the position—although it does best in full sun.

How to grow

Once planted and established, the tree needs no regular attention. Pruning should not be necessary. Caraganas prefer a light soil, but are adaptable.

Propagation

Seed is the easiest for an amateur. Chip and presoak in tepid water for 12 hours, then sow in

Caragana arborescens 'Pendula'

individual pots in warmth in spring. Germination usually takes about a month. Grow indoors for another month, then harden off and stand outdoors until planted in the autumn.

Cuttings are best left to commercial growers.

SOME POPULAR SPECIES	
C. arborescens *(Siberia)* An erect, vigorous and sparsely-branched shrub, often trained to a tree form, with pinnate leaves	with four to six pairs of light green leaflets. Yellow, pea-type flowers in late spring. Large shrub or small tree.

Caryopteris

Blue spiraea

SH/AK/FS

The name is derived from the Greek *karnon* (nut) and *pteron* (wing), referring to the winged fruit. It is a genus of low-growing deciduous shrubs needing protection from frost in the first winter.

Caryopteris x *clandonensis*

How to grow

Although caryopteris thrive in alkaline (chalk) soils, they will grow well in most soils. Unfortunately they are not always hardy in exposed areas. This is not much of a problem, as even though the tops get cut back new shoots should appear in spring. If not cut back by frost, prune to two or three buds of the old wood in early spring.

Propagation

Seed is not often used to propagate caryopteris, and the species below and its varieties should be raised from cuttings.

Cuttings of semi-ripe wood can be taken in mid summer and rooted in a cold frame. Pot up in spring if rooted, and place outdoors in a sunny spot after hardening off. Plant in autumn.

SOME POPULAR SPECIES	
C. x clandonensis *(Garden origin)* Low-growing shrub with aromatic, narrow, dull green leaves and clusters of bright blue tubular	flowers during late summer and early autumn. Varieties include 'Arthur Simmonds' (bright blue, and hardy) and 'Heavenly Blue'. Small.

Catalpa
Indian bean tree
TR/O/PS–FS

A genus of 11 species of hardy deciduous trees, although their large leaves are easily damaged by wind in an exposed area, and they generally do best in mild areas.

How to grow
They do best in moist, but not too heavy soil, in full sun. But once established they will need no routine care. Pruning should not be necessary.

Propagation
Seed-raised plants are slow to reach planting size. Cuttings are a more sensible method of propagation.

Take cuttings in mid summer, using semi-ripe wood, and root in a propagator. Pot up singly and overwinter in a cold frame. Plant out in a nursery bed in spring, and let the plant grow on for a couple of years before planting out.

Catalpa bignonioides 'Aurea'

SOME POPULAR SPECIES	
C. bignonioides *(Eastern North America)* Large, broadly ovate leaves, heart-shaped at the base ('Aurea' has large, pale yellow leaves). Large	heads of white flowers marked yellow and purple in mid summer. Sometimes followed by long, green 'beans', turning black later. Large.

Ceanothus (evergreen)
Californian lilac
SH(E)/O/FS

The ceanothus are not ideal plants for a very cold or exposed site, but are very desirable shrubs well worth growing in more favourable areas. The deciduous species (see page 22) are generally hardier than the evergreens listed here. It is best to treat ceanothus as wall shrubs, where they benefit from the protection of the wall.

How to grow
Always plant ceanothus in spring, to give them a chance to become established before winter. Well-drained soil and a sunny site will give the best chances of success. The evergreen species need little or no pruning, although they can be trimmed to shape after flowering.

Propagation
As for deciduous species, page 22.

SOME POPULAR SPECIES	
C. dentatus *(California)* Erect, bushy shrub. Small blue flowers in late spring and summer. Large. **C. impressus** *(California)* Rather tender, twiggy	shrub. Deep blue flowers in spring. Large. **C. thyrsiflorus** Upright habit. Pale blue flowers. One of the hardiest evergreen species.

Ceanothus impressus

Ceanothus (deciduous)

Californian lilac

SH/O/FS

The evergreen species were described on page 21. The deciduous ceanothus are rather hardier, generally have larger leaves and looser flowers.

How to grow

The deciduous species should have flowering shoots cut back to within 7.5 cm (3 in) of the previous year's growth in early spring.

Propagation

Seed can be used for some of the species, although cuttings are the most popular method, and must be used for hybrids and varieties.

Take semi-ripe cuttings in mid summer, and use a propagator. Pot up singly when rooted, usually after about a month, and overwinter in a cold frame. Harden off in the spring, and plant out in the autumn.

Ceanothus 'Gloire de Versailles'

SOME POPULAR SPECIES	
C. x delilianus *(Garden origin)* Panicles of blue flowers throughout the summer. It is usually offered in one of	the selected clones, such as 'Gloire de Versailles' (powder blue), and 'Topaz' (light indigo blue). Medium.

Ceratostigma

Hardy plumbago

SH/O/FS

Although the common name implies that this is a dependably hardy plant, it can be unreliable in cold districts. In a cold winter the stems may be killed by frost even in more favourable areas, but this does not matter because new stems are likely to emerge in the spring.

How to grow

Give the plant a good start by providing a sunny spot, away from cold winds, and preferably on well-drained soil. Plant in groups to make a bold display. Once it is established, prune hard, almost to ground level, in mid spring.

Propagation

Cuttings of semi-ripe wood are best taken in mid summer and rooted in a propagator. Pot up singly and move to a cold frame to overwinter. Harden off and plant out in late spring.

Suckers can be separated carefully and lifted in mid spring. Replant well-rooted portions immediately.

SOME POPULAR SPECIES	
C. willmottianum *(Western China)* Stalkless dark green diamond-shaped leaves (turning red in autumn).	Clusters of bright blue phlox-like flowers nestle among the leaves from mid summer to mid autumn. Small.

Ceratostigma willmottianum

Cercidiphyllum

Katsura tree

TR/O/FS

A genus of only two deciduous trees, *C. japonicum* being the one most commonly seen.

How to grow
Plant in a sunny, sheltered position. Best on a deep, fertile soil, and in districts with a favourable climate, though it will grow elsewhere. No pruning or other regular care should be necessary.

Propagation
Seed can be sown in containers under a cold frame in early spring, barely covering with compost. Keep seedlings in the frame until the following spring, when they can be planted in a nursery bed for a couple of years.

Layers should be ready to lift in about 18 to 24 months. Peg down shoots two or three years old in spring.

SOME POPULAR SPECIES	
C. japonicum *(Japan)* Large tree with branches slightly pendulous at their tips. Rounded, toothed, green leaves, tinted red as	they unfold. In autumn they turn red and yellow. In very cold areas the new leaves may be damaged by early frosts.

Cercidiphyllum japonicum

Cercis

Judas tree

TR/O/FS

A genus of seven species of deciduous trees native to China, North America and southern Europe. The blooms are carried on naked stems.

How to grow
Cercis are not very hardy trees, and in cold districts they are best grown in a sheltered position, perhaps near a wall. They need good soil and full sun.

Propagation
Seed should be presoaked in tepid water for 24 hours before sowing in containers in a cold frame or outdoors in early or mid winter (to allow a period of chilling). In spring, place in a propagator—germination takes about a month. Pot seedlings singly and keep in a cold frame for the summer and winter. Then pot on into 13 cm (5 in) pots in spring, harden off and stand outdoors from late spring. Plant in the autumn.

Layers can be pegged down in autumn or spring, using shoots two or three years old. Remove a sliver of bark on the underside. Plant out after 18 to 24 months.

SOME POPULAR SPECIES	
C. siliquastrum *(Orient, Southern Europe)* Wide-spreading tree, often bushy in habit. Roundish leaves with	heart-shaped bases. Bright purplish-rose pea-shaped flowers are borne on bare branches in late spring. Medium.

Cercis siliquastrum

Chaenomeles

Flowering quince, japonica

SH/O/PS–FS

To many older gardeners the chaenomeles are still regarded as cydonias (the true quinces), in which genus they were formerly included. The fruits that they produce are edible, though not very palatable and are usually used in preserves.

How to grow

Choose a suitable site. They will grow almost anywhere, but unless carefully positioned will look straggly and untidy. They are best growing up against a wall, or as a loose, informal hedge.

Grown as a bush, the only pruning necessary will be to thin out overcrowded branches after flowering. As a wall shrub, cut back the previous season's growth to two or three buds after flowering.

Propagation

Seeds are used mainly to increase the species— the named varieties and hybrids will have to be propagated by one of the vegetative methods that follow. Sow in containers in a cold frame in autumn, barely covering with compost. Prick out seedlings into individual pots in spring and over-winter under the frame. Harden off and plant out in a nursery bed to grow on for 18–24 months before planting.

Layers are easy. Make a slicing cut half way through the underside of shoots two or three years old, and peg down in autumn. They should be ready to plant in about two years.

Cuttings can be taken from semi-ripe wood in

Chaenomeles x *superba*

summer and rooted in a propagator. Pot up after rooting (usually it takes about a month) and overwinter under frames. Harden off in spring and grow in a nursery bed for two years before planting out.

Grafting is widely used by commercial growers. It is done in early spring using seedling rootstocks.

SOME POPULAR SPECIES	
C. japonica *(Cydonia japonica) (Japan)* Wide-spreading and relatively low-growing shrub with ovate to round mid green leaves and bright orange-flame apple-blossom-type flowers in clusters of two or three in spring. Apple-like yellowish fruits in late summer. Small. **C. speciosa** *(C. lagenaria, Cydonia speciosa) (China)* Well-branched, spreading	shrub. Dark green glossy leaves and bowl-shaped flowers about 5 cm (2 in) across from mid winter to mid spring. Many varieties are available, in shades of red or pink, as well as white. Small to medium. **C. x superba** *(Garden origin)* Vigorous, free-flowering hybrid. Many varieties, flowering in spring, in shades of red, orange, and pink. Small to medium.

Chimonanthus

Winter sweet

SH/O/FS

The name comes from the Greek *cheima* (winter) and *anthos* (flower), which gives an indication of the plant's main attribute: the ability to flower in the middle of winter, regardless of the weather. It also has the bonus of fragrance, hence its com-

mon name. It is, however, to be viewed at close quarters, as it will not make much of a display.

How to grow

For best results, plant near a west-facing or south-facing wall. You may need patience, as the flowers are reluctant to appear until the plant is well established and reasonably mature.

No regular pruning is necessary, but remove unwanted or damaged branches after flowering. But wall-grown specimens should have shoots

Chimonanthus praecox

that have flowered cut back to within a few inches of the older growth.

Propagation
Seed-raised plants are slow to flower. Named varieties will have to be propagated vegetatively anyway. Layering is easier and more reliable.

Layer in early autumn, using shoots two or three years old and removing a sliver of wood from the underside. They should be ready to plant in about two years.

SOME POPULAR SPECIES	
C. praecox *(C. fragrans)* *(China)* Bushy shrub with willow-like mid green leaves. Flowers appear on naked branches in winter. These	are about 2.5 cm (1 in) across and very fragrant. There are improved forms such as 'Grandiflorus' (deeper yellow with a red stain). Medium.

Choisya
Mexican orange blossom
SH(E)/O/PS–FS

A genus of six evergreen flowering shrubs,

Choisya ternata

named in honour of a Swiss botanist, Jacques Denys Choisy. Only one species is grown.

How to grow
In favourable districts this shrub is hardy, but it will be affected by extreme weather and in cold areas it is best grown as a wall shrub. It should thrive on any well-drained soil, and will do best in full sun.

No pruning is normally required, but cut back any frost-damaged branches in spring.

Propagation
Cuttings root easily. Root them in a propagator in mid or late summer, using semi-ripe wood. In about a month they should have rooted and be ready to pot individually to overwinter in a cold frame. Harden off and stand outdoors for the summer, ready to plant in the autumn.

SOME POPULAR SPECIES	
C. ternata *(Mexico)* Rounded shrub with glossy, stalkless leaves, arranged in threes, fragrant when crushed. Sweetly-scented white,	five-petalled flowers in mid and late spring; a few flowers may appear spasmodically until autumn. Medium.

Cistus

Sun rose, rock rose

SH(E)/O/FS

A genus of evergreen shrubs from southern Europe, of varying hardiness. Although the individual flowers are fleeting, they are produced abundantly in long succession. In Britain those listed should do well in southern areas, but even there they may be killed in a severe winter. For that reason it is always worth taking some cuttings each year to replace casualties.

How to grow

Cistus will not tolerate heavy soil (they do best on poor, well-drained ground) or too much shade. Avoid frost pockets, and protect the plants from cold winds.

Although regular pruning is not necessary, frost-damaged or unwanted branches should be removed. Avoid cutting into old wood.

Propagation

Seed is a useful method of propagation for the species (hybrids and varieties will have to be raised from cuttings). Sow in containers in a cold frame in early spring. Prick out singly into 8 cm (3 in) pots, and later move to 13 cm (5 in) pots. Overwinter in a cold frame. Harden off and plant out in late spring.

Root cuttings of semi-ripe wood in a propagator in mid or late summer. They should root in about a month, when they can be potted up for overwintering in a cold frame. Pot on in spring, then harden off and stand outdoors for the summer. Move back to the frame for the winter months, and then plant out in late spring the following year.

Top *Cistus* x *purpureus*
Bottom *C.* x *cyprius*

SOME POPULAR SPECIES			
C. x corbariensis *(Garden origin)* Dense, bushy, spreading shrub. Dull green wavy-edged leaves. 4 cm (1½ in) white flowers with yellow stain at the base of each petal. Flowers late spring and early summer. Small.	**C. x cyprius** *(Garden origin)* Olive green, sticky leaves. White flowers, 7.5 cm (3 in) wide with crimson-maroon blotches at the base of the petals, early and mid summer. Medium shrub.	**C. ladanifer** *(South West Europe)* An erect species. White flowers up to 10 cm (4 in) across, with chocolate blotches. Small. **C. x lusitanicus** *(Garden origin)* White flowers about 5 cm	(2 in) wide with crimson blotches, early summer. **C. x purpureus** *(Garden origin)* Vigorous, upright bush. Rosy-crimson flowers with chocolate basal blotches, late spring to mid summer. Small.

Clematis (large-flowered type)

CL/AK/FS

This is a large genus with about 250 species, some of them tender. Most, but not all, are climbers. The most popular kinds are the large-flowered climbing hybrids and varieties. The small-flowered species are described in the next entry.

How to grow
Clematis thrive on alkaline (chalky) soil, but good results can be obtained on most fertile soils. And although they like to flower in full sun, the plants will do better if the roots are cool and shaded (which is what happens in nature when a climber scrambles up through a tree).

Pruning clematis can be something of a nightmare—it varies according to the type. Spring-flowering varieties that bloom on growth made the previous year are pruned immediately after flowering; summer-flowering varieties in spring. Those that flower in late summer or autumn, in the leaf axils of the current season's growth, are pruned back hard in early spring—to within a few inches of the previous season's growth. This is a simplified explanation, as the method of training and the natural vigour of the plant also need to be considered. Most good catalogues indicate the type of pruning required for the particular varieties offered.

Mulch the plants each spring.

Top *Clematis* 'Hagley Hybrid'
Bottom *C.* 'Jackmanii Superba'

Propagation
Seed is not used for the hybrids and large-flowered varieties.

Cuttings are taken in mid summer. Use internodal cuttings (see illustration on page 107) and root in a propagator. After about a month, pot up singly and overwinter in a cold frame. In mid spring, pot on into 13 cm (5 in) pots and harden off in late spring to stand outside until autumn, when they can be planted.

Layers are easy, and should be pegged down in spring. Select vigorous shoots of the previous season's growth. Kink the stem mid-way between two leaf joints, at the point of pegging, to improve rooting. The layers should be ready for planting out a year later.

SOME POPULAR SPECIES	
Garden hybrids Some good varieties include 'Ernest Markham' (carmine-red, early summer to early autumn); 'Hagley Hybrid' (pink with chocolate-brown anthers, early summer to early autumn); 'Jackmanii Superba' (violet-purple, mid summer to early autumn); 'Lasurstern' (deep lavender-blue, late	spring, early summer); 'Nelly Moser' (pale mauve-pink with carmine bars, late spring, early summer); 'Ville de Lyon' (bright carmine-red, deeper at margins, mid summer to autumn); and 'Vyvyan Pennell' (double violet-blue, late spring to mid summer—single flowers may be produced in autumn).

Clematis (species)

CL/AK/FS

The large-flowered clematis were included on page 27, the species below generally have smaller flowers but are usually easier to grow and have better 'covering power' than the large-flowered types.

How to grow

The general advice for the large-flowered hybrids (page 27) applies to the species too, though unless you have to confine the plant in some way pruning should be unnecessary. Many of them, such as *C. montana*, will scramble happily through a tree or over a fence.

Clematis montana 'Rubens'

Propagation

Although cuttings and layers (see large-flowered hybrids) are equally easy and successful for the species, you can also raise species from seed.

Sow in containers in a cold frame in spring. Prick out into individual pots and harden off to stand outdoors from early summer until autumn, when they can be planted.

SOME POPULAR SPECIES

C. alpina *(South and Central Europe)* A frail-looking plant. Dark green leaves formed of nine lance-shaped, coarsely toothed leaflets. Pendulous violet-blue flowers in mid and late spring. Moderate vigour.

C. macropetala *(Siberia and Northern China)* Bushy, low-growing climber with violet-purple bell-like nodding flowers with paler centre, in late spring and early summer. There is a pink variety. Moderate vigour.

C. montana *(Himalayas)* A vigorous and reliable climber. Dark green trifoliate leaves and 5 cm (2 in) wide four-petalled white flowers in late spring. Several good forms are available, including 'Rubens' (pink flowers, bronze-purple leaves). Vigorous.

C. orientalis *(Manchuria, China, Himalayas, Iran, Caucasus)* (Orange-peel clematis) Much-branched, sprawling climber with deeply lobed and somewhat fern-like leaves. Nodding, slightly fragrant, yellow flowers from late summer to mid autumn. Moderate to vigorous.

C. tangutica *(China)* Finely divided grey-green leaves and yellow lantern-like nodding flowers about 5 cm (2 in) across, from late summer to mid autumn. Silky seed heads. Moderate vigour.

Clerodendrum

SH/O/FS

A diverse genus of about 400 species of trees, shrubs and climbers from Africa and Asia, but there are only two of them that we grow outdoors, plants very different in appearance and habit. You may find the genus spelt clerodendron.

How to grow

Plant in fertile soil in full sun, preferably in a sheltered spot. The top growth of *C. bungei* is likely to be cut down by frost, but should shoot again from the base in spring.

Clerodendrum trichotomum fargesii

Pruning should not be necessary, other than cutting back frost-damaged shoots to sound wood.

Propagation

Seed can be germinated in a propagator in spring after prechilling or stratifying. Grow on the seedlings in individual pots under cover, and pot on into 13 cm (5 in) pots the following spring. Harden off and stand outside for the summer. Protect with frames for the second winter, then plant out in the spring.

Cuttings of semi-ripe wood can be taken in mid or late summer. These should be 10–13 cm (4–5 in) long, with a heel and rooted in a propagator. Overwinter in individual pots in a cold frame and harden off to plant out in a nursery bed for the following summer. Grow on for 12–18 months before planting.

Suckers provide the easiest means of propagation. Detach them in early spring or mid autumn and grow on in a nursery bed for a year, then plant in their permanent positions.

SOME POPULAR SPECIES	
C. bungei *(Eastern China, Japan)* Suckering shrub with large heart-shaped, leaves that smell unpleasant when crushed. Ball-like heads of star-like rosy-red flowers are produced in late summer and early autumn. This species is rather tender and requires a sheltered position. Medium.	**C. trichotomum** *(Eastern China, Japan)* Bushy, but sparse slow-growing shrub with ovate or oval mid green leaves, which have an unpleasant odour when crushed. Lax, erect heads of scented star-like pinkish-white flowers in late summer or early autumn, followed by blue berries set in crimson calyces. *C. t. fargesii* fruits more freely. Large.

Clethra

SH/AD/PS–FS

A genus of evergreen or deciduous small trees and shrubs, notable for their fragrant flowers. All require lime-free soil.

Clethra alnifolia

How to grow

Do not try to grow them on chalky soil. They will tolerate damp ground, and need plenty of humus in the soil. A lightly shaded position suits them best.

Pruning is not normally necessary.

Propagation

The plants are usually raised from cuttings or layers. Use semi-ripe wood with a heel for the cuttings, and root in a propagator in mid summer. Overwinter in a cold frame before planting out in late spring. Use a lime-free compost.

Shoots two or three years old can be layered in early autumn. They should be ready for planting in about two years.

SOME POPULAR SPECIES	
C. alnifolia *(Eastern USA)* (Sweet pepper bush) Erect, bushy shrub with fragrant 5–15 cm (2–6 in) long clusters of creamy white flowers in late summer to mid autumn. 'Paniculata' has larger flowers. Medium	**C. fargesii** *(China)* Fragrant pure white flowers carried in terminal clusters 15–25 cm (6–10 in) long in mid summer. The leaves turn yellow in autumn. A very beautiful large shrub or small tree.

Colutea

Bladder senna

SH/O/FS

Colutea arborescens

A genus of 26 species, though only one is in general cultivation, its inflated seed pods being the main feature.

How to grow

Thrives in poor, sandy ground, but will tolerate heavy soils too. Needs full sun.

The shrub is inclined to become leggy, in which case the shoots can be shortened to half their length in early spring.

Propagation

Seed can be sown in warmth in spring, after presoaking for 24 hours and chipping. Sow in individual pots and barely cover with compost. Overwinter in a frame. Harden off to plant out in spring.

Cuttings can be difficult to root and slow to establish. Use 10 cm (4 in) cuttings of ripe wood with a heel, in late summer. Use a rooting hormone. Once rooted, pot into 10 cm (4 in) pots to overwinter in a frame. Harden off in spring and stand outdoors until autumn when they can be planted.

SOME POPULAR SPECIES	
C. arborescens *(Southern Europe, Mediterranean areas)* Strong-growing, branching shrub with	pinnate leaves. Yellow pea-like flowers in summer, followed by inflated bladder-like seed pods. Large.

Cornus (shrubby species)

Dogwood, cornel

SH/O/PS–FS

The dogwoods or cornels range from low shrubs to trees. Some of them are grown for flowers, others for foliage, some for their coloured stems. The species that assume a tree form are included in the next entry. The ones below are grown as shrubs. Botanists have reclassified some of the dogwoods into new genera, such as *Swida* and *Cynoxylon*, but these are not names you need to worry about because in catalogues and garden centres *Cornus* is the name that you will find these shrubs listed under.

Cornus alba 'Sibirica'

How to grow
Best in moist soil. The flowering types need no pruning, but those grown for a crop of coloured stems should be cut back to within a few inches of ground level in early spring.

Propagation
Seed is a very slow method and you will be better with one of the following alternatives (if you want to try seed, see tree species).

Cuttings are normally used to increase hybrids and named varieties. Take semi-ripe heel cuttings in mid or late summer, and root in a propagator. Pot up and overwinter in a cold frame. Harden off in spring before planting out in a nursery bed for two or three years.

Suckers provide a quick and easy method of propagation. Detach in late autumn and replant immediately in a nursery bed. Grow on for a year before planting out.

Layers of young growth should be pegged down in early autumn, after slitting the stems on the underside. They should be ready to plant out a year or two later.

SOME POPULAR SPECIES			
C. alba *(North Korea, Siberia to Manchuria)* (Red-barked dogwood) Vigorous, upright, suckering shrub with oval to round mid green leaves, turning red or orange in	autumn. Thicket of upright stems conspicuous in winter. Desirable varieties include: 'Elegantissima' (red stems, leaves edged and mottled white), 'Sibirica'	(crimson stems), and 'Spaethii' (red stems, leaves variegated). Medium. **C. florida** *(Eastern USA)* (Flowering dogwood) Wide-spreading, much-branched shrub. Dark	green leaves, pale beneath, turning orange and scarlet in autumn. Small green flowers surrounded by petal-like white bracts in late spring. 'Rubra' has pinkish bracts. Large.

Cornus (tree species)

Dogwood, cornel

TR/O/FS

The shrubby dogwoods were included in the previous entry, but there are also some fine trees, many of them holding their branches in elegant tiers, some with masses of bloom (actually the bracts are usually the conspicuous part), and sometimes interesting fruit.

How to grow
Given a fertile soil and full sun, all these trees should be trouble-free and need no regular attention such as pruning. Some, such as *C. nuttallii*, are not suitable for shallow chalk soils.

Propagation
Seeds are slow to produce results but are sometimes used for the true species. Sow in containers in a cold frame in early autumn. They may take up to 18 months to germinate. Pot up the seedlings singly and overwinter in the frame. Harden off in spring and grow on in a nursery bed for a couple of years.

Cuttings and layers are easier—see shrubby species.

SOME POPULAR SPECIES	
C. kousa *(Japan, Korea)* Shrub or small tree with shrubby habit. Ovate and slender-pointed mid green leaves turning bronze and crimson in autumn. *C. k. chinensis* has more open growth, slightly larger leaves, and very fine autumn colour. Small or very small.	**C. mas** *(Europe)* (Cornelian cherry) Shrub or tree with twiggy growth and an open, bushy habit. The clusters of yellow flowers appear on bare branches from late winter to mid spring. Occasionally red berries 2 cm (¾ in) long are produced. Small.

Cornus kousa chinensis

Corylopsis

SH/AD/PS–FS

A genus of 20 hardy, deciduous early-flowering shrubs or small trees. Grown for the fragrant, bell-shaped flowers in catkin-like racemes.

How to grow
These shrubs dislike an alkaline soil rather than needing an acid one. Provided the soil is well enriched with humus, they will grow on a wide range of soils. Because they come into growth early, they will appreciate the protection of a sheltered position, perhaps by a warm wall, or in open woodland.

Propagation
Cuttings are shy-rooting, and layers are easier. If you want to take cuttings, take them in mid summer. Make them 10 cm (4 in) long and use semi-ripe wood with a heel. Use a rooting hormone and a propagator. Overwinter in a cold frame, then harden off in spring to plant out in a nursery bed for a couple of years.

   Layers are best pegged down in mid autumn. Choose shoots two or three years old and slit the underside of the stem. They are usually ready for planting out within two years.

Corylopsis willmottiae

SOME POPULAR SPECIES	
C. pauciflora *(Japan)* Spreading shrub with erect branches and slender twiggy growth. Pointed, ovate, bright green leaves, heart-shaped at their bases. Pale yellow, fragrant, drooping racemes in early and mid spring. Needs a sheltered position for the best display. The most tolerant of chalk. Medium.	**C. spicata** *(Japan)* Similar to previous species, but greenish-yellow flowers in slightly longer racemes. Medium. **C. willmottiae** *(China)* Upright, with an open habit. Roundish to oval leaves, with heart-shaped bases. Sweetly-scented soft yellow flowers in drooping racemes. Medium to large.

Corylus

Hazel, cobnut

SH–TR/O/PS–FS

A genus of 15 species of trees and shrubs, some of them long cultivated for their nuts. There are, however, some useful ornamental species and varieties. The name of the genus is thought to come from the Greek *korys* (a helmet or hood), in reference to the calyx on the nuts.

How to grow
Hazels do best in fertile soil, although you are unlikely to be growing them for the nuts and they should accommodate themselves to most soils.

   Pruning is unlikely to be necessary for those grown as ornamental trees or shrubs.

Corylus avellana 'Contorta'

Propagation

Seed can be sown in containers in a cold frame (but protect the nuts from vermin!). Pot up in spring when the seedlings are large enough to handle. Overwinter in the frame and harden off to plant in a nursery bed in spring, where they can remain for a year or two until planting out.

Layers are a reliable and easy way to propagate hazels. Peg down young shoots in late autumn after removing a sliver of bark on the underside. Layers should be ready to move to a nursery bed within a year. Grow on as for seedlings.

Suckers can be lifted in autumn, selecting those that are well rooted. Plant out in a nursery bed. This is an easy way to propagate a few plants. Grafting is little used, but can be carried out in early spring.

SOME POPULAR SPECIES	
C. avellana *(North Africa, Western Asia, Europe)* Small tree or thicket-forming erect shrub. Serrated, broadly oval mid green leaves with heart-shaped bases. Yellow catkins in late winter, sometimes followed by clusters of roundish nuts. Good varieties: 'Aurea' has yellow leaves, 'Contorta' (the corkscrew hazel) has very unusual twisted branches and is slow growing. Large shrub or small tree.	**C. colurna** *(Western Asia, South-eastern Europe)* (Turkish hazel) Dark green, broadly ovate leaves. Grey, corrugated bark is a feature in winter. Yellow catkins in late winter or spring, followed by nuts. Large tree. **C. maxima** *(Western Asia, Southern Europe)* (Filbert) Robust shrub (sometimes a tree). Catkins in late winter, followed by nuts. 'Purpurea' has purple leaves. Large.

Cotinus

Smoke tree, Venetian sumach
SH/O/FS

A genus of three deciduous shrubs and trees, previously included in *Rhus*. The plant most widely grown is *C. coggygria*, an off-putting name for a very desirable plant, commonly known as the smoke tree, because its plume-like inflorescences carried above the foliage look like a smoky haze on a sunny day.

Cotinus coggygria

How to grow

Although any soil is suitable, these shrubs do better on light soils. Good drainage is essential. Avoid a very rich soil, otherwise the autumn colour may not be so good.

Although good border plants, these shrubs are perhaps best seen planted as isolated specimens.

Propagation

Seed is slow to produce worthwhile plants. It is better to use cuttings or root cuttings.

Root cuttings are taken in winter. They should be about 4–5 cm (1½–2 in) long and rooted in a propagator. Pot up when growth is seen and grow on under frames for the next winter. Grow on in a nursery bed for another year before planting.

Cuttings are relatively easy. Take 13 cm (5 in) heel cuttings of ripe wood in late summer. Root in a cold frame, and pot up into 10–13 cm (4–5 in) pots the following spring. Harden off and stand outside for the summer. Give winter protection again, then plant out in spring.

SOME POPULAR SPECIES	
C. coggygria *(Rhus cotinus) (Europe to the Caucasus)* Rounded, bushy, and often spreading shrub. Light green leaves colouring brilliantly in autumn. Loose, feathery	heads of small purple flowers in mid summer. 'Atropurpureus' *(C. c. purpureus)* has purplish-pink flowers. 'Foliis Purpureis' has dark plum-purple foliage. Medium to large.

Cotoneaster (evergreen)

SH(E)/O/PS–FS

There are about 50 species of cotoneaster, both deciduous and evergreen. Deciduous species are described in the following entry. Both types are greatly valued for their conspicuous, usually red, berries.

How to grow
Cotoneasters are easy to grow in any ordinary soil, but are best in full sun. Pruning is not normally required although vigorous specimens can be pruned back hard in mid spring when they become too large.

Propagation
Seed of the straight species (not varieties) can be sown in containers under a cold frame in mid autumn. Prick out the seedlings when they are large enough to handle—it may take 6–18 months to germinate and reach this stage. Overwinter under the frame, then harden off and plant out in a nursery bed in late spring. Grow on for a couple of years before setting out in their final positions.

Cuttings are more commonly used than seed,

Cotoneaster salicifolia

especially for hybrids and named varieties. See deciduous species.

Layers are an easy method of propagation. Peg down stems two or three years old in mid or late autumn, after slicing half way through from the underside. Young plants are usually rooted and ready for lifting in a year. Grow on for a year or two before planting out.

SOME POPULAR SPECIES

C. conspicuus *(South East Tibet)*
Mound-forming shrub with arching stems with small mid green shining leaves. Masses of white flowers in early summer, bright red berries in autumn. *C. c.* 'Decorus' is small and suitable for a rock garden. Small to medium.

C. dammeri *(China)*
Prostrate shrub with creeping stems. Dark green, glossy, oval leaves. White flowers in early summer; followed by round, eventually coral-red berries. Good for clothing banks. Prostrate.

C. franchetii *(China)*
Semi-evergreen. Slender, arching shrub. Sage green leaves, whitish beneath. Pinkish flowers in early summer; orange-scarlet berries in autumn. Medium.

C. frigidus *(Himalayas)*
Semi-evergreen. Rounded shrub or small tree with dull green leaves pale and woolly beneath. Clusters of small white flowers in early summer; pea-sized bright red berries that last well into winter. The variety 'Fructuluteo' has large bunches of creamy-yellow berries in autumn and winter. Large shrub.

C. microphyllus *(Himalayas)*
Dwarf, wide-spreading shrub with dark green, oval, glossy leaves. White flowers in late spring and early summer. Round, scarlet-red berries crowd the stems later in the year. Ideal for covering banks. Prostrate.

C. 'Hybridus Pendulus' *(Garden origin)*
Carpeting shrub with arching branches, perhaps most attractive grown as a standard (grafted on to a stem of *C. frigidus*) when it forms a weeping head with branches reaching the soil. On its own roots is ideal for covering banks.

Clusters of white flowers in early summer; brilliant red, round berries in autumn. Prostrate; very small tree (as standard).

C. salicifolius *(China)*
Willow-like glossy leaves, hairy and greyish-white beneath. White flowers in early summer, followed in autumn by bright red berries. Several excellent varieties are available, including 'Autumn Fire', sometimes sold as 'Herbstfeuer' (semi-evergreen, lax habit, masses of orange-red fruits), and 'Repens' (very narrow leaves and red fruits). Large (the species) to dwarf ('Repens').

Cotoneaster (deciduous)

SH/O/PS–FS

Evergreen species were described in the previous entry. The deciduous species make up for lack of year-round cover by having good autumn colour. The berries will often keep the shrubs colourful until mid winter.

How to grow

As evergreen species, but any pruning is best done in late winter.

Propagation

Seed can be used for the normal species (see evergreens) but hybrids and varieties are likely to differ from their parents.

Cuttings are the main method. Root semi-ripe wood cuttings in a cold frame in mid summer, making the sections 13–15 cm (5–6 in) long. For evergreens, make the cuttings 10 cm (4 in) long and take them in late summer or early autumn. Pot up rooted cuttings in spring, harden off and plant out into a nursery bed in late spring. They should be ready for final planting in one or two years.

SOME POPULAR SPECIES	
C. bullatus *(Western China)* Attractive shrub with a few long and arching shoots and a rather sparse, lax habit. Corrugated, oval, glossy leaves. Small pinkish-white flowers in early summer followed by brilliant red berries; good autumn leaf colour. Large.	**C. horizontalis** *(China)* (Fish-bone cotoneaster) Distinctive, fish-bone arrangement of stems, crowded with small dark green leaves. Small pink flowers in late spring; red berries in autumn. Spreads out horizontally above the soil, but can be trained up a wall where it can reach 1.8 m (6 ft).

Cotoneaster horizontalis in berry

C. horizontalis in flower

Crataegus
Hawthorn, May
TR/O/PS–FS

A genus of 200 species of deciduous and generally very tough trees and shrubs. The species listed below are very tough and will thrive on poor soil, on chalk, and in exposed sites. They make neat, small trees.

How to grow
Although they will tolerate partial shade, the hawthorns will do better in full sun. No regular pruning or other care is necessary.

Propagation
Seed is a popular method for the species. Stratify the seeds outdoors (see page 117) for 18 months before sowing in a container under a cold frame in late winter or early spring. Grow the seedlings on under the frame until the following spring, then harden off and prick out into a nursery bed to be grown on for a couple of years.

Special forms are budded on to seedling rootstocks, using the 'T' method in mid summer, but this is seldom attempted by amateurs.

Grafting, using the whip-and-tongue method, is used for some varieties. It is done in early spring, but is not an easy method of propagation.

Crataegus x *lavallei*

SOME POPULAR SPECIES			
C. crus-galli *(Eastern North America)* (Cockspur thorn) Flat-topped tree with spreading branches displaying rigid, viciously sharp thorns and tapering pear-shaped to ovate glossy green leaves, which colour well in autumn. Clusters of white flowers in early summer, followed by persistent red haws. Small to medium	**C. x lavallei** *(Garden origin)* Sturdy, leafy tree with densely-branched habit. Deeply cut and downy dark green leaves that persist on the tree well into early winter. Clusters of white flowers in early summer. Orange-reds fruits with brown speckles in autumn. Medium. **C. monogyna** *(Europe)* (Common hawthorn) Densely packed	branches and glossy dark green three- to seven-lobed leaves. Fragrant clusters of white flowers in late spring or early summer. Masses of red haws during autumn. Varieties include 'Biflora' (the Glastonbury thorn; may produce some flowers in winter), 'Pendula Rosea' (pink flowers on pendulous branches), and 'Stricta' (upright, erect branches). Large.	**C. oxyacantha** *(C. laevigata) (Europe)* (Hawthorn) Thorny tree with lobed glossy green leaves and highly scented white flowers in late spring. Roundish, crimson-red fruits in autumn. Good varieties include 'Paul's Scarlet', also sold as 'Coccinea Plena' (double scarlet flowers), 'Plena' (double white flowers), 'Rosea' (single pink flowers). Medium.

Cytisus
Broom
SH/O/FS

The brooms are sun-loving shrubs, most of them spectacular in flower in late spring and early summer. There are many hybrids that add to the range of plants available.

How to grow
Cytisus flourish in poor, sandy soil. Most will tolerate chalk, but some of the hybrids may be short-lived on shallow chalky soils. As brooms resent disturbance, always buy pot-grown plants and do not move them about once planted.

C. scoparius and its hybrids tend to become leggy with age. You can help to delay this by pruning annually once they are established, cutting off most of the previous season's growth, but being careful not to cut into old wood.

Other species mentioned can have those stems that have borne flowers cut back by about half their length after flowering (again being careful not to cut into old wood).

C. battandieri is best treated as a wall shrub: after flowering cut out old, stray shoots and weak wood, then tie in the best of the new shoots.

Propagation
Seed is suitable for the true species, but hybrids and varieties are likely to be variable (this may not matter). Presoak the seeds in tepid water for 12 hours before sowing in a propagator in early or mid spring. Prick out when large enough to

Cytisus scoparius 'Canary Bird'

handle (usually after four to six weeks), and harden off to stand outdoors from early summer onwards. Plant in their final positions in autumn.

Cuttings are used for named varieties, but they can be shy rooting. Take 10 cm (4 in) heel cuttings of semi-ripe wood and root in a cold frame in mid summer. Pot up singly when rooted in spring. Harden off and plunge outdoors from late spring onwards, then plant in early autumn.

SOME POPULAR SPECIES			
C. battandieri *(Morocco)* Upright shrub with silvery leaves formed of three leaflets. Eye-catching pineapple-scented golden-yellow cone-shaped flower clusters in late spring and early summer. Best against a warm high wall where it will get some protection from frost. Large.	**C. x beanii** *(Garden origin)* Semi-prostrate shrub with spreading but rounded habit. Golden-yellow flower sprays in late spring. Ideal for a rock garden. Dwarf. **C. x kewensis** *(Garden origin)* Low, spreading shrub with masses of cream flowers in late spring. Suitable for a rock garden. Dwarf.	**C. x praecox** *(Garden origin)* (Warminster broom) Vigorous, bushy growth and masses of sulphur-yellow flowers in mid or late spring. Varieties include 'Albus' (white), and the lovely 'Allgold' (arching sprays of bright yellow long-lasting flowers). Small to medium.	**C. scoparius** *(Western Europe)* (Common broom) Upright-growing shrub with bright green branches and trifoliate mid green leaves. Yellow pea-type flowers in late spring and early winter. There are many varieties, mainly in shades of yellow or red, sometimes with two colours. Medium.

Daboecia

Connemara heath, Dabeoc's heath

SH(E)/AD/FS

A small genus of evergreen sub-shrubs—the only dependably hardy one being *D. cantabrica.* Daboecias associate well with heathers, being of similar habit and requiring the same conditions.

How to grow

A well-drained, acid soil is necessary for good results. Add plenty of peat when planting and mulch with peat annually. Cut off the dead flowered shoots with shears.

Propagation

Seed is seldom used for propagation, but can be tried for fun (see Calluna, page 18).

Layers are fairly easy. In early or mid spring, peg down stems one or two years old, about 15–20 cm (6–8 in) from the tips, after first twisting and kinking the stems. Cover the pegged wounded area with lime-free compost. The young plants should be ready for lifting after about a year. Pot into 10 cm (4 in) pots and grow on for another year, or plant out if well rooted.

Division is possible in spring or autumn. Treat as rooted layers.

Daboecia cantabrica

SOME POPULAR SPECIES
D. cantabrica *(Iberian Peninsula, Ireland, France)* Slender stems covered with pointed glossy dark green leaves. Small purple-pink bell-like flowers all summer and into autumn. Varieties include 'Alba' (white), and 'Atropurpurea' (dark rose-purple). Dwarf to small.

Daphne

SH(E, some)/O/PS–FS

A genus of 70 evergreen and deciduous shrubs from Europe and Asia. Those below are hardy and fragrant.

How to grow

Daphnes need a lot of humus, so add plenty of peat or compost when planting, and mulch each spring. Pruning should not be necessary, cut out straggly shoots in early spring.

Propagation

Seed is a slow method of propagation. Sow in containers under a cold frame in mid autumn. Pot up seedlings singly into 8 cm (3 in) pots when

Daphne mezereum

ready to handle (usually in 12 to 24 months). Harden off in spring and plunge outdoors for the summer. Bring back under the frame for winter protection. Pot on in spring, then grow on for a second summer and plant out in autumn.

Cuttings are more reliable. Use sections of semi-ripe wood about 5–8 cm (2–3 in) long, preferably with a heel, in mid or late summer. Root under a cold frame, and pot up rooted cuttings in spring, using 8 cm (3 in) pots. Harden off and treat as for seedlings.

Layers are easy. Peg down suitable shoots in autumn. They should be ready for planting out in 18 to 24 months.

SOME POPULAR SPECIES			
D. x burkwoodii *(Garden origin)* Semi-evergreen shrub with dense, rounded shape. Fragrant, long-tubed soft pink flowers in late spring or early summer. Small. **D. cneorum** *(Central and Southern Europe)* (Garland flower) Near-	prostrate evergreens with trailing shoots and narrowly oblong deep green leaves. Highly scented terminal clusters of rose-pink flowers in late spring and early summer. Varieties include 'Alba' (white) and 'Eximia' (crimson buds opening rose-pink). Prostrate.	**D. mezereum** *(Asia Minor, Siberia, Europe)* (Mezereon) Erectly-branched deciduous shrub with eye-catching clusters of fragrant purple-red flowers on bare stems in late winter and early spring. The scarlet berries are poisonous. There are several varieties, including	'Grandiflora', which starts to flower from early autumn. Small. **D. odora** *(China, Japan)* Crowded terminal heads of reddish-purple very fragrant flowers from mid winter to mid spring. 'Aureomarginata' has leaves edged creamy-white. Medium.

Deutzia

SH/O/PS–FS

Deciduous, early-flowering shrubs with an abundance of flowers, named after Johann van der Deutz, a friend and patron of Thunberg, who named the genus.

How to grow
Some deutzias are vulnerable to late spring frosts, so avoid an open or exposed position, especially if the garden is in a low-lying area.

Prune just after flowering, removing over-crowded or very old shoots at ground level. Other shoots are best left untouched.

Propagation
Cuttings are the normal method of propagation. Take 8 cm (3 in) semi-ripe cuttings in mid summer and root under a cold frame. Pot up rooted cuttings individually in spring. Harden off and plant out in a nursery bed in late spring.

Alternatively 25 cm (10 in) hardwood cuttings can be rooted under a cold frame in late autumn. They should be ready to plant out into a nursery bed the following autumn, where they will need to grow on for another year.

Deutzia x *elegantissima* 'Fasciculata'

SOME POPULAR SPECIES	
D. x elegantissima *(Garden origin)* Upright, bushy habit. Fragrant, rose-tinted, star-shaped flowers in panicles on arching branches, late spring and early summer. Small to medium. **D. x rosea** *(Garden origin)* Arching branches with rounded clusters of pink	bell-shaped flowers in early summer. Small. **D. scabra** *(China, Japan)* Erectly-branched shrub. Large panicles of white flowers in early and mid summer. Varieties include 'Plena' (double white, tinted rose-purple outside) and 'Pride of Rochester' (double white, tinted pink). Medium.

Elaeagnus

SH(E, some)/O/PS–FS

A genus of about 45 shrubs grown mainly for their foliage. Some of the variegated ones are among the most striking foliage plants in the garden, especially in winter.

How to grow

These plants will grow well on most soils, including chalk. Deciduous species prefer full sun; the evergreens will grow well in partial shade but look better, and the variegated kinds will have more impact, in sun.

Pruning should not be necessary, but unwanted branches can be cut back in spring.

Propagation

Seed is used mainly for deciduous species. Sow in early autumn in containers under a cold frame, barely covering the seed. Prick out seedlings singly into 8 cm (3 in) pots as soon as large enough to handle in spring or summer. Overwinter under frames. Harden off in spring and plant out in a nursery bed to grow on for a year or two.

Cuttings are the main method of propagation for evergreen species. Take 8 cm (3 in) semi-ripe cuttings in mid summer and root in a cold frame. In spring pot up rooted cuttings individually into 10 cm (4 in) pots. Harden off and stand outdoors during the summer; they should be ready for planting in the autumn.

Top *Elaeagnus* x *ebbingei*
Bottom *E. pungens* 'Maculata'

SOME POPULAR SPECIES

E. angustifolia *(Southern Europe, Western Asia)* (Oleaster) Deciduous shrub or small tree. Branches occasionally spined. Willow-like silvery grey-green leaves. Silvery flowers with bell-shaped tubes, in early summer, followed by yellowish-silver fruits. Large.
E. commutata *(E. argentea) (North America)* (Silver berry) Slow-growing suckering shrub. Erect habit and slender branches with silvery-green oval leaves. Profusion of fragrant silvery flowers in late spring, followed by egg-shaped fruits. Medium to large.
E. x ebbingei *(Garden origin)* Fast-growing evergreen with silvery-grey leathery leaves. Silvery flowers in mid and late autumn. Orange, silver-speckled egg-shaped fruits may be produced in spring. Useful seaside plant. Large.
E. pungens *(Japan)* Spreading, vigorous evergreen with leathery, oval or oblong glossy leaves (duller when young). Grown mainly in the form 'Maculata' (glossy green leaves with bold central splash of gold). Large.
E. umbellata *(China, Korea, Japan)* Wide-spreading and often thorny deciduous shrub. Narrowly oval bright green leaves, silvery beneath. Clusters of funnel-shaped silvery flowers, creamy-white inside, in late spring. Round, silvery then red fruits. Large.

Enkianthus

SH/AD/PS

The name comes from the words *enkuos* (swollen) and *anthos* (flower), and presumably refers to the pouch at the base of the corolla of some species—the flowers are like small nodding bells.

How to grow
An acid soil is preferred, although you may succeed on a neutral soil if you add plenty of peat when planting, and mulch with peat each spring. Plant in a sheltered position. Pruning should not be necessary.

Propagation
Seed should be surface-sown in containers of lime-free compost and stood outside in early and mid winter, then put in a warm place—16°C (60°F)—to germinate after about six weeks. The seeds need light to germinate well, so do not cover. Pot up when the seedlings can be handled (usually after about six weeks in warmth), and grow on under frames. Pot on in spring, harden off and plunge outdoors during the summer. Give frame protection during the winter, then plant out the following spring.

Cuttings are shy-rooting. Use semi-ripe wood 8 cm (3 in) long, preferably with a heel, in late summer. Root under a cold frame in lime-free compost and pot up in spring in the same mix. Harden off and plant out in a nursery bed in late spring. Grow there for two or three years, until ready to be planted.

SOME POPULAR SPECIES	
E. campanulatus *(Japan)* Upright shrub, occasionally a tree, with clustered pear-shaped oval dull green leaves, assuming brilliant colours in autumn. Pendulous clusters of bell-shaped creamy-	yellow flowers with red veins in late spring. Large shrub. **E. perulatus** *(Japan)* Leafy, slow-growing shrub. White flowers in spring. Leaves colour brilliant red in autumn. Medium.

Top *Enkianthus campanulatus*
Bottom *E. perulatus*

Erica

Heath, heather

SH(E)/AD/PS–FS

There are over 500 species in the world, most of them natives of South Africa. Only a few species are grown as garden plants, but there is a vast number of varieties and hybrids of these.

How to grow

E. herbacea (still well known as *E. carnea*), *E.* x *darleyensis* and *E. mediterranea* (which botanists now prefer to put into other species), will grow on chalky soil, provided it is not too shallow. Add peat when planting, and mulch with peat each spring.

Clip off dead flower stems as soon as the flowers have faded (if you find the dead heads decorative, leave them on for the winter and trim off in spring).

Propagation

Seed can be sown in a greenhouse in a lime-free compost in mid or late winter—see Calluna, page 18, but cuttings provide a much more reliable method of propagation.

Root new season's tip cuttings from late spring to mid summer in a propagator or under a cold frame. Use a lime-free compost and treat as Callunas, page 18.

Layers are a good method if you want only a few plants. Peg down vigorous young growths about 20 cm (8 in) from the tips in early or mid spring. First bend or kink the stems. Peg down the wounded area and cover with 5 cm (2 in) of lime-free potting compost, leaving the tips exposed. They should be ready to lift and replant in 12 months.

Erica cinerea 'Atrosanguinea'

SOME POPULAR SPECIES

E. cinerea *(Western Europe)*
(Bell heather) A low-growing shrub. Bright purple flowers on wiry stems from early summer to late autumn, though colour and flowering time depend on the variety. There are many of them, a few have golden foliage. Dwarf.

E. x darleyensis *(Garden origin)*
A hybrid between *E. herbacea* and *E. mediterranea*. Tolerates lime well, but not suitable for *shallow* chalk soils. There are several good varieties, including 'Arthur Johnson' (magenta), 'George Rendall' (pink), and 'Silberschmelze' (white). Dwarf.

E. herbacea *(E. carnea) (Central Europe)*
This plant is still widely sold as *E. carnea*, but the new name is becoming increasingly well used. Low, tufted habit, becoming prostrate and spreading with age. The original species has rosy-red flowers, it is the many varieties that are grown. These span flowering from late autumn to mid spring. Some have golden foliage. Prostrate.

E. mediterranea *(E. erigena, E. hibernica) (Ireland, France, Spain)*
Although the correct name is now *E. erigena*, this plant is most likely to be found under its older names. Dense, bushy shrub with erect and glabrous branches, and narrow dark green leaves in whorls. The type has rosy-red flowers from late winter to late spring. But there are varieties with flowers in shades of pink, red, and white. Will tolerate lime if the soil is not too shallow or dry. May be vulnerable in cold areas. Medium.

E. tetralix *(Northern and Western Europe)*
(Cross-leaved heath) Low-growing shrub that spreads with age and becomes prostrate. Narrow, grey-green, hairy leaves. Pink flowers in summer. There are several varieties, mainly with pink flowers and grey leaves. Prostrate.

E. vagans *(Western Europe)*
(Cornish heath) Low-spreading shrub. Pinkish-purple flowers from mid summer to mid autumn. There are many varieties mainly in shades of pink, red, and white. Dwarf to small.

Escallonia

SH(E)/O/FS

A genus of about 60 species of evergreen and deciduous shrubs or small trees from South America, named in honour of Señor Escallon, a Spanish traveller in South America. The species widely grown, and their hybrids, are evergreen. Unfortunately they are of borderline hardiness in cold districts, where they are best treated as wall shrubs. In milder areas they make excellent hedges.

How to grow

Any well-drained soil should suit escallonias, but they do best in a warm, sunny position. The protection of a warm, sunny wall will be useful in cold areas.

In autumn, prune back shoots that have flowered.

Propagation

Cuttings of firm, semi-ripe wood are taken in mid or late summer. Root under a cold frame or (preferably) in a propagator. They should root in about a month in good conditions. Pot up and overwinter under a cold frame. Pot on in mid spring and harden off and plant out in a nursery bed in early summer. Grow on for one or two years then plant in their permanent positions.

SOME POPULAR SPECIES
E. macrantha *(South America)* Dense green bush with small, glossy leaves. A particularly good coastal shrub. Small rose-crimson flowers in summer. A parent of many of the hybrids. Medium. **Garden hybrids** Most of the escallonias grown are hybrids. They all have small evergreen leaves and small cup-shaped flowers. Popular varieties include 'Apple Blossom' (pale pink), 'C. F. Ball' (crimson), 'Donard Seedling' (white), and 'Peach Blossom' (pink). These are medium-sized.

Escallonias are among the most decorative of flowering evergreen shrubs

Eucalyptus

Gum tree

TR(E)/O/FS

A large genus of over 500 species of mainly tall evergreen trees. Although growing rapidly into tall trees, some popular species, such as *E. gunnii* can be cut back each year to make bushy plants producing juvenile foliage. The juvenile foliage is generally round, sometimes clasping the stem; foliage on mature plants is more elongated.

Eucalyptus perriniana

How to grow

Well-drained soil in a sunny position suit the eucalypts. Hardiness depends on the species, but even the hardiest may be cut back in a very severe winter—a setback from which many will recover. The more tender species will benefit from a screen or other protection for the first winter.

Pruning is only necessary if the tree is being 'coppiced' for juvenile foliage. Cut the young plant back to a few inches above the ground in spring. Then each spring cut the stems back.

Propagation

Easily raised from seed. Sow in warmth under glass (preferably in a propagator) in early or mid spring. Sow two or three seeds to an 8 cm (3 in) pot, then thin to one seedling after germination.

The plants can stand outdoors or in a frame for the summer after hardening off, but overwinter under glass and plant out in late spring after hardening off.

SOME POPULAR SPECIES	
E. gunnii *(Tasmania)* (Cider gum) Young leaves are rounded and vary from blue-green to silvery-white. Gradually, after the second year, the adult dark blue-green, lance-shaped leaves appear. Large to very large.	**E. perriniana** *(New South Wales, Victoria, Tasmania)* (Round-leaved snow gum) Silvery leaves, paired and rounded in juvenile form, clasping the stem— making it attractive for stooling annually. White bark with dark blotches.

Euonymus (evergreen)

SH–CL(E)/O/PS–FS

This large genus contains many useful garden shrubs, diverse in use and appearance. Deciduous species are described in the next entry. The two evergreens included here are among the most desirable and adaptable evergreen foliage shrubs.

How to grow

The variegated varieties are best in full sun but they are not fussy about soil and will cope with chalk or acid, good or indifferent.

No pruning is usually necessary, but they can be cut back hard in late spring if required.

Euonymus fortunei 'Silver Queen'

Propagation

Seeds can be used (treat as for deciduous species) but the evergreen species are more often increased by cuttings.

Cuttings of semi-ripe wood can be taken in mid or late summer, preferably with a heel, and rooted in a cold frame. They should be ready to pot up in about a month. Overwinter in the frame, then harden off in spring and in late spring plant out in a nursery bed or pot on into larger containers. Either way, grow on for a year or two until they are large enough to plant out.

SOME POPULAR SPECIES			
E. fortunei *(E. radicans)* *(Japan)* Creeping, prostrate shrub that can also be trained against a wall, where it will grow to 2.4 m (8 ft) or so high. Allowed to scramble over the soil it will form a large mat of growth. Young plants usually have	the prostrate tendency, with more erect shoots coming later. Glossy green leaves. Popular varieties include: 'Emerald Gaiety' (silver and green variegation), 'Emerald 'n' Gold' (green and gold variegation), and 'Silver Queen' (creamy white and	green variegation, slow-growing). Usually prostrate. **E. japonicus** *(Japan)* Upright, bushy, shrub with dense growth. Narrowly oval glossy, dark green leaves. As a shrub (rather than a hedge) is usually grown in one of the	variegated forms, such as: 'Aureopictus' (central gold splash), 'Ovatus Aureus' (leaves edged creamy yellow), and 'Microphyllus Variegatus' (small leaves, white margin). Large (some varieties small or medium).

Euonymus (deciduous)

SH/O/FS

Evergreen species were included in the previous entry. The deciduous species lack the year-round interest of the evergreens, but they have rich autumn colours. They are also good on chalk.

How to grow
As evergreen species.

Propagation
Seed can be sown in a loam-based compost in early or mid autumn. Germinate in a cold frame or cool greenhouse—but be prepared to wait up to 18 months. Prick the seedlings out into small pots as soon as they appear. Overwinter under cover, harden off in spring, and plant out in a nursery bed in late spring—or pot on. Either way, grow on outdoors for another one or two years before final planting.

For just a few plants you will find cuttings a quicker method—see evergreen species.

Euonymus alatus

SOME POPULAR SPECIES	
E. alatus *(China, Japan)* Slow-growing, open-centred shrub with stiff habit. Corky wings on branches. Dark green leaves colour brilliantly in autumn. Medium. **E. europaeus** *(Europe)* (Common spindle tree) Spreading, bushy shrub or small tree. Narrowly oval, sometimes pear-shaped, green leaves. Small greenish-white flowers in late spring. Red fruits	open to display orange seed coats in autumn. There are several varieties. Medium. **E. yedoensis** *(Japan, Korea)* Sturdy, flat-topped shrub or small tree with stiff branches. Brilliant red and yellow autumn colour. White flowers in late spring followed by rose-pink fruits that remain on the plant after the leaves have fallen. Large.

Fagus
Beech
TR/O/SD–FS

There are 10 *Fagus* species, but only *F. sylvatica* is widely planted. It will tolerate most soils, including chalk, will grow in cold, windswept areas, and put up with atmospheric pollution. Despite this it is not a tree to plant unless you have a large garden. Shallow rooting and the dense shade cast, also mean that underplanting is very difficult.

How to grow
Beech will tolerate any soil that is not wet or heavy. No routine care is necessary.

Propagation
Seed is slow, but the only way to increase green-leaved beech. In mid autumn, soak the seeds in tepid water for about 12 hours, and sow in containers. Stand outdoors to chill (protect from vermin), and in late winter move into a cold frame or into slight warmth. Germination is erratic and may take from a month to a year-and-a-half. Prick out individual seedlings as they become large enough to handle, and overwinter in the cold frame. Harden off then plant in a nursery bed to grow on for two or three years.

Fagus sylvatica 'Purpurea Pendula'

Grafting is used for the varieties, but it is a difficult procedure needing a supply of three-year-old rootstock seedlings.

SOME POPULAR SPECIES	
F. sylvatica *(Europe)* The ordinary beech is a magnificent tree, but more at home in a park. If you have space in the garden, try a variety such as	'Purpurea Pendula' (small mushroom-headed purple-leaved weeping tree) or 'Dawyck' (also known as 'Fastigiata', a tall but columnar tree).

Forsythia
Golden bell bush
SH/O/PS–FS

William Forsyth, an eighteenth-century superintendent of the Royal Gardens at Kensington, London, gave his name to a genus containing some of the most popular flowering shrubs. They are among the most brilliant in flower, and are all the more useful for flowering early in the season.

How to grow
Although forsythias will thrive in partial shade, they will be even better in full sun.

Forsythia x *intermedia* 'Spectabilis'

Prune carefully, making sure you do it immediately after flowering. Remove any very old or damaged shoots first, then shorten vigorous shoots to keep the plant tidy. You can remove shoots that have borne flowers at this time, but be careful not to remove new growth later in the season otherwise you may be removing next-year's flowers too.

Propagation

Seed is not normally used—if you want to try, stratify or prechill the seed and sow in warmth in spring.

Cuttings are the most reliable method. Take 10–13 cm (4–5 in) semi-ripe cuttings in mid summer and root under a cold frame. Pot up into 9 cm (3½ in) pots in early autumn, and overwinter under the frame. Harden off in spring, and plant in a nursery bed to grow on for another year.

Alternatively, root 25–30 cm (10–12 in) hardwood cuttings under a cold frame in late autumn. Lift the plants a year later.

Layers of *F. suspensa* are easy. Slit a suitable young stem and peg down in mid autumn. They are usually ready to plant out a year later.

SOME POPULAR SPECIES	
F. x intermedia *(Garden origin)* Compact, vigorous shrub with masses of yellow flowers in early and mid spring. The varieties usually grown include 'Spectabilis' (profusion of bright yellow flowers), and 'Lynwood' (masses of large, rich yellow flowers). Medium.	**F. ovata** *(Korea)* Bushy, often spreading shrub. Ovate, dull green leaves. Amber-yellow flowers in early spring or late winter. Small. **F. suspensa** *(China)* Rambling shrub, suitable for training against a wall. Pendulous yellow flowers in clusters of two or three in spring. Medium.

Fothergilla

SH/AD/PS–FS

A genus of four hardy deciduous shrubs from North America. Dr Fothergill (1712–80), after whom the genus was named, grew many American plants in his Essex garden in England. They are grown mainly for their autumn leaf colour.

How to grow

Fothergillas need a lime-free soil, and preferably one with a high humus content. Add peat when planting. No regular pruning required.

Propagation

Seed is not the usual method of propagation, but you can try sowing it in a container in a cold frame in mid autumn, barely covering the seed. Prick out seedlings as they appear (it might take a year or even longer), and overwinter in the cold frame. Pot on in spring, harden off and plunge outdoors for the summer, ready to plant in autumn. Use a lime-free compost at all stages.

Layering is a quicker and more satisfactory method. Peg down branches two or three years old in early or mid autumn. Remove a sliver of bark on the underside of the stem first. The layers should have rooted and be ready to lift and replant in about two years.

SOME POPULAR SPECIES	
F. major *(USA)* Rounded habit, glossy, dark green summer leaves turning orange, yellow, and red during autumn. Spikes of white flowers in late spring, before the	leaves have fully developed. Medium. **F. monticola** *(USA)* Similar to above, and considered by some to be a form of it. More spreading habit. Medium.

Fothergilla monticola

Fuchsia

SH/O/PS–FS

A genus of about 100 species, named in honour of Leonard Fuchs, a German botanist. Few of the species are grown, but there are hundreds of popular hybrids. Most of them are tender, though some of the popular large-flowered hybrids will overwinter outdoors if protected. The species below are fairly hardy and will survive most winters in favourable areas. If they are cut down to ground level they should sprout again in spring.

How to grow
Plant deeply and cover the crowns with peat, to provide additional winter protection. They will grow best in well-drained soil with plenty of humus, so add garden compost or peat when planting, and mulch each spring.

Cut down the plants in late autumn or early winter and cover the roots with bracken or a layer of peat if you are in a less than favourable area. In mild districts you can just cut back the stems to 2.5 cm (1 in) above the ground in early spring.

It is worth applying a general fertiliser a couple of times during the growing season.

Propagation
Seed is occasionally used for hardy species. It should be germinated in warmth, hardened off and planted out in summer.

Fuchsia magellanica

Cuttings are more dependable. Take 8 cm (3 in) tip cuttings in early or mid spring, and root under glass at about 16°C (60°F). They should root within a month. Pot up into 9 cm (3½ in) pots. Grow on and overwinter in a frame, then harden off and plant out in early summer.

SOME POPULAR SPECIES	
F. magellanica *(Mexico, Chile, Peru)* One of the hardiest fuchsias and widely grown outdoors in mild areas, where it is sometimes used as a hedging plant. The pendent 4–5 cm (1½–2 in) long crimson and purple flowers last from mid summer until the growth is cut down by frost. Varieties include *F.*	*m. gracilis* (more slender habit, scarlet and white flowers), 'Riccartonii' (deeper calyx and broader sepals), and 'Versicolor' (variegated). Small to medium. **Garden hybrids** Many of the large-flowered hybrids are suitable for favourable areas—consult a specialist catalogue.

Garrya
Silk tassel bush
SH(E)/O/PS–FS

A genus of 18 species of hardy and tender evergreen shrubs, named to honour Nicholas Garry of the Hudson's Bay Company, who helped David Douglas on his plant collecting expeditions. Few species are hardy enough to be grown in Britain, and these need the benefit of a warm wall in all except the most favourable areas. Buy or propagate male plants, as these have the best catkins.

Garrya elliptica

How to grow
Buy a container-grown plant as garryas do not transplant easily, and plant in spring to give it a chance to settle before the winter. Some form of protection, such as a polythene screen, is advisable for the first winter. Although garryas will tolerate partial shade, they will do better in full sun.

Pruning is not normally necessary, but straggly shoots can be trimmed back in mid or late spring.

Propagation
Cuttings of semi-ripe wood are usually taken in mid or late summer and rooted in a cold frame. In early or mid spring, pot up the rooted cuttings into 9 cm (3½ in) pots. Harden off and stand outdoors in early summer. Grow on in pots until early autumn when they should be ready to plant out in their final positions. Do not let the plants dry out at any time.

SOME POPULAR SPECIES	
G. elliptica *(California, Oregon)* Fast-growing, vigorous bushy shrub. Dark green, shiny, leathery leaves. Pendulous catkins, often 15 cm (6 in) long in the	case of males, in late winter and early spring. Female catkins smaller and silver-grey. 'James Roof' is a male with extra-long catkins. Good wall shrub. Medium to large.

Gaultheria

SH(E)/AD/PS

A genus of 200 species of evergreen flowering trees and shrubs, grown for their evergreen cover and attractive berries. Unfortunately they are likely to be disappointing unless you have an acid soil or plant them in a peat bed.

How to grow
A moist, acid soil is essential. Although partial shade is desirable, don't plant where drips from trees will be a problem.

Although no regular pruning is needed, *G. shallon* can be rather invasive if conditions suit it, and it may need cutting back hard in mid or late spring to control growth.

Propagation
Seed can be surface sown in containers of lime-free compost in a cold frame in mid autumn. Move into a warm greenhouse (preferably a propagator) in late winter. They should be ready for pricking out in one to two months. Set two or three seedlings together in small pots. Gradually reduce the temperature and move to a cold frame after a couple of weeks. Harden off and stand outdoors for the summer, returning to a frame for the winter. Plant out in a nursery bed in mid spring and grow on there for a couple of years.

Cuttings are the usual method. Prepare semi-ripe heel cuttings in mid or late summer and root in a cold frame. In spring pot up the rooted cuttings, harden off and stand outdoors until autumn, when they can be planted. Use a lime-free compost for all stages.

The easiest method of propagating *G. shallon* is to remove suckers. They will grow away readily if severed and replanted in early or mid autumn.

SOME POPULAR SPECIES	
G. procumbens *(North America)* (Checkerberry, Partridge-berry, Winter-green) Low, spreading, somewhat tufted shrub with slender stems. Thick, leathery, dark green leaves. White or pink flowers in mid or late	summer. Round, bright red berries. Prostrate. **G. shallon** *(Western North America)* Suckering shrub. Pinkish-white flowers carried in late spring and early summer; clusters of dark purple fruits. Small to medium.

Gaultheria procumbens

Genista

Broom

SH/O/FS

A genus of deciduous or almost leafless shrubs, grown for the profusion of pea-type flowers produced in late spring or early summer. They actually do well on poor soil.

Genista lydia

How to grow

Although genistas do not transplant well (buy container-grown plants), once established they will thrive in any well-drained soil if given a sunny spot.

It is worth pinching out the growing tips of young plants to encourage a bushy habit.

Regular pruning is not necessary, but you can cut back stems that have flowered to maintain an open shrub. Do not cut back into old wood.

Propagation

Seeds can be used for the true species, but cuttings should be used for hybrids and forms that do not come true from seed. Sow indoors in a propagator in early spring after soaking for 12 hours and chipping. Prick out singly into 9 cm (3½ in) pots as soon as large enough to handle. Harden off and stand outdoors for the summer. Plant in their final positions in autumn.

Cuttings are used for hybrids and varieties, though they can be shy to root. Use semi-ripe wood, preferably with a heel, in mid or late summer and root in a cold frame. They should be rooted and ready to pot up in spring. Harden off and treat as for seedlings.

SOME POPULAR SPECIES	
G. aetnensis *(Sardinia, Sicily)* (Mt Etna broom) Loose, open habit. Pale, rush-like branches. Golden-yellow flowers in summer. Large. **G. hispanica** *(South West Europe)* (Spanish gorse) Dense, spiny, much-branched shrub. Deep green narrowly lance-shaped	leaves. Pea-type yellow flowers in early and mid summer. Small. **G. lydia** *(Syria, Southern Europe)* Lax habit with pendulous, spreading shoots. Bright yellow pea-type flowers in late spring and early summer. Excellent for trailing over low walls or for covering banks. Small.

Gleditsia

Honey locust

TR/O/PS–FS

You may sometimes see this genus spelt Gleditschia (it was named after an eighteenth-century director of Berlin Botanic Garden, J. Gottlieb Gleditsch). The genus contains species of deciduous trees with pea-type flowers and attractive pinnate leaves. The common name of honey locust is due to the fact that the seed pods of some species are lined inside with a sweetish pulp, in which the seeds are embedded. They are good town trees.

Gleditsia triacanthos 'Sunburst'

How to grow
Little attention and no routine pruning should be necessary. Well-drained soil will give the best results.

Propagation
Seed should be sown indoors in spring, first presoaking for 24 hours in tepid water and then chipping. If given 18–21°C (65–70°F), they should germinate in 15 to 30 days. Prick out into 9 cm (3½ in) pots and overwinter under a cold frame. Pot on in spring, harden off, and plunge the pots outdoors for another year or two before planting.

SOME POPULAR SPECIES	
G. triacanthos *(North America)* Light green pinnate leaves becoming clear yellow in autumn. Spines on mature	trunks and branches. 'Sunburst' (spineless stems, golden foliage) is good. Medium to large.

Hamamelis

Witch hazel
SH/AD/PS–FS

A genus of six species of hardy deciduous shrubs, useful because most species flower in winter on the bare branches. The common name witch hazel is reputed to stem from the name given to *H. virginiana* by early settlers of North America, because of the similarity of habit and leaf shape to that of the British hazel, which was used in divining for water.

How to grow
Best on an acid or neutral, well-drained soil. Incorporate plenty of garden compost or peat at planting time. Avoid an exposed or wind-swept position. Pruning should only be necessary to remove damaged or unwanted branches.

Propagation
Seed can be sown in containers of lime-free compost under a cold frame in autumn, but germination is slow and may take about two years. Harden off after overwintering in a frame and plant out in a nursery bed in late spring, to be grown on for two years.

Cuttings have a low success rate, so use a rooting hormone. Take 10 cm (4 in) hardwood cuttings with a heel in early autumn, and root in a propagator. They should then root in about six to eight weeks. Pot up without delay and overwinter in a frame. Harden off in spring and plant out as for seedlings. Use a lime-free compost.

Layering is the easiest method. Slit and peg down branches two or three years old in early autumn. They should be ready to lift and plant in a nursery bed in two years. Grow on for another two years before planting in their final positions.

Grafting is possible, but is really a job for the professionals.

SOME POPULAR SPECIES	
H. x intermedia *(Garden origin)* Ovate to pear-shaped mid green leaves, with good autumn colour. Yellow or copper-tinted petals, crimped and twisted, are borne on bare branches in late winter. There are several varieties. Medium. **H. japonica** *(Japan)* (Japanese witch hazel) Sparsely-branched, spreading shrub, similar to above species. 'Arborea' is a good variety with almost	horizontal branches and abundant flowers. Medium to large. **H. mollis** *(Central China)* (Chinese witch hazel) Spreading branches, often ascending at the tips. Roundish to pear-shaped mid green leaves that turn yellow in autumn. Fragrant, golden-yellow flowers, flushed red at their base, in thick clusters on bare branches in mid winter. 'Pallida' is a good variety. Medium.

Hamamelis mollis

Hebe

SH(E)/O/FS

A genus of about 100 species, formerly included in *Veronica*, and sometimes called shrubby veronicas. Most species grown are hardy only in mild areas or in coastal regions. You may find that they will survive a normal winter but be killed by an exceptionally severe one. They are such desirable shrubs, however, that it is well worth the risk of occasional losses.

How to grow

Hebes will grow in most soils, and are satisfactory on chalk. They need well-drained soil and full sun. The more tender species will need a sheltered position, perhaps by a wall in inland gardens.

Regular pruning is unnecessary, but cut back straggly growth in later spring when any frost-damaged branches are removed.

Propagation

Seed is not generally used. But if you want to try, sow under glass in spring (16–18°C/60–65°F). Pot up seedlings individually and overwinter under a cold frame. Harden off and stand outdoors in early summer to grow on until autumn, when they can be planted.

Cuttings are the normal method of propagation. Take 5–8 cm (2–3 in) semi-ripe cuttings in mid or late summer and root under a cold frame. Pot up into 8–9 cm (3–3½ in) pots in early or mid spring and harden off before treating as seed-raised plants.

Hebe pinguifolia 'Pagei'

SOME POPULAR SPECIES			
H. anomala *(New Zealand)* Slow-growing, compact shrub with crowded small bright green leaves. White or pale pink flowers in clustered spikes at the ends of shoots in early and mid summer. One of the hardiest. Small. **H. armstrongii** *(New Zealand)* Small, much-branched shrub with clusters of twigs that often spread out like the rays of the sun.	Very small deep golden-yellow leaves and white flower spikes. The plant grown in gardens under this name may be *H. ochracea.* Small. **H. brachysiphon** *(H. traversii) (New Zealand)* Rounded, bushy, often spreading shrub with crowded narrowly ovate dark green leaves. White flowers from leaf axils near the ends of the shoots in early and mid summer. Small to medium.	**H. elliptica** *(New Zealand, Chile, Tierra del Fuego, Falkland Islands)* Compact, dome-shaped shrub with pale to mid green leaves. Fragrant white flowers in clusters near ends of the shoots in summer. Small. **H. x franciscana** *(Garden origin)* Compact shrub with green leaves and dense racemes of bright blue flowers, which appear throughout the summer. One of the	hardiest hebes. 'Variegata' is less vigorous, but has leaves margined creamy-white. Flowers mauve. Dwarf to small. **H. pinguifolia 'Pagei'** *(H. 'Pagei', H. pageana) (New Zealand)* Mat-forming. Glaucous grey leaves. Small white flowers in late spring and early summer. Excellent shrub for ground cover. Prostrate.

Hedera

Ivy

CL(E)/O/SD–FS

The ivies need little introduction, but it is not always appreciated that they produce two forms of growth—juvenile leaves are lobed, while adult or arborescent growth has entire leaves. The arborescent growth does not produce aerial roots, but it does carry the flowers and fruit. The different growth forms explain the sometimes different appearance of the same species or variety.

How to grow

These are tough plants and once established should need no further care other than hacking back when they outgrow their bounds.

Variegated varieties are best in a position in good light.

Propagation

Cuttings of semi-ripe tips—10 cm (4 in) long—are taken in mid or late summer. If a large number of plants is needed, make leaf bud cuttings. Root under a cold frame—the cuttings should be ready for potting up individually in 8 cm (3 in) pots in about a month. Grow on and overwinter in the frame, then harden off and plant out in spring.

Layers of climbing and trailing varieties root easily if stems are pegged down in autumn. Young plants should be ready to lift and plant out in a year or less.

Top *Hedera helix* 'Goldheart'
Bottom *H. h.* 'Glacier'

SOME POPULAR SPECIES

H. canariensis *(North Africa, Canary Islands)* (Canary Island Ivy) Vigorous, fast-growing climber with large, leathery bright green leaves, broadly ovate with heart-shaped base. In winter they frequently become bronze-green. The form usually grown is 'Variegata' (often sold as 'Gloire de Marengo'); the leaves have a dark green centre, variegated with silver-grey towards the edges. May be killed in severely cold weather. Moderate to large climber.

H. colchica *(Caucasus, Southern Anatolia)* Fast-growing climber with ovate to heart-shaped glossy green leaves. The two most usually grown forms are 'Dentata' (very large leaves) and 'Dentata Variegata' (margined creamy-yellow maturing and fading to creamy-white). Moderate to large climber.

H. helix *(Europe)* Exceptionally hardy. It is usually the more decorative varieties that are grown. Popular ones include 'Buttercup' (yellow, becoming greenish with age), 'Chicago' (small, dark green often blotched bronze-purple), 'Glacier' (silvery-grey, white edges), 'Goldheart' (central splash of yellow), and 'Hibernica' (large dark green leaves; grows fast and is good as a ground cover). Most are vigorous.

Helianthemum

Sun rose, rock rose

SH/O/FS

The name of this genus comes from the Greek *helios* (sun) and *anthemon* (flower). This, and the common name of sun rose, should tell you that these plants like a hot, sunny position. Given the right conditions they will flower profusely, and spread vigorously too.

How to grow

Well-drained soil and a sunny site are the essentials for these plants. Heavy soil and shade mean disappointment. They also do better on poor soil than on rich soil.

H. nummularium and its varieties will have to be cut back hard with shears after flowering to keep them neat and within bounds. Remove the dead flowers—it may induce further flowering later in the year.

Propagation

Seed is not widely used, but can be sown under glass in early spring, preferably in a propagator. Seedlings are usually ready to prick out into individual pots in about a month. Grow on and overwinter under a cold frame. Harden off and plant out in mid or late spring.

Cuttings are the normal method of propagation. Prepare 8 cm (3 in) heel cuttings of non-flowering shoots in summer, and root in a cold frame. They should have rooted and be ready for potting up in about a month. Treat as described for seedlings.

SOME POPULAR SPECIES	
H. nummularium *(Europe)* The most widely grown species, usually in the form of its many varieties. Low-growing, semi-shrubby plant with lax, scrambling stems. Elliptic deep green leaves, paler	beneath. Masses of flowers in early and mid summer. Varieties include 'Ben Afflick' (orange and buff flowers), 'Ben Heckla' (bronze-gold), 'Jubilee' (yellow), and 'Wisley Pink' (pink). Prostrate.

Helianthemum nummularium

Hibiscus

Tree hollyhock

SH/O/FS

Hibiscus is an ancient Greek name used by Dioscorides for the marsh mallow. In fact it is a very diverse genus and only one species is widely used as a hardy shrub. It is one of the finest late-flowering shrubs.

How to grow

Well-drained, fertile soil and full sun are the basic requirements. In cold areas it is best to plant by a wall.

Little pruning is necessary. Just cut back branches that are too long, after flowering.

Propagation

Seed is not often used as cuttings are more reliable and are sure to produce plants like the parents if you are growing one of the varieties.

Prepare 8 cm (3 in) semi-ripe heel cuttings in mid or late summer. Root under glass, preferably in a propagator, and move into 8 cm (3 in) pots when rooted—usually after about a month. Overwinter in a cold frame, pot on in mid spring,

and plunge outdoors for the summer after hardening off. Plant out in autumn.

SOME POPULAR SPECIES	
H. syriacus *(Syria)* Bushy shrub with erect branches. Three-lobed, coarsely-toothed leaves. Trumpet-mouthed flowers, about 7.5 cm (3 in) wide, from late summer to mid autumn.	Varieties include 'Blue Bird' (violet-blue, dark eye), 'Hambro' (bluish, crimson eye), 'Red Heart' (white, red centre), and 'Woodbridge' (pink, carmine eye). Medium.

Hibiscus syriacus 'Woodbridge'

Hippophae
Sea buckthorn
SH/O/PS–FS

A small genus of hardy trees and shrubs grown mainly for their attractive berries. The most popular species, *H. rhamnoides* is, as the common name suggests, a good seaside plant, where it will make an excellent windbreak. You will need male and female plants for berries (which are generally shunned by birds because they contain an acrid juice).

How to grow
Although ideal for a sandy soil in a coastal area, hippophae will grow in any well-drained ordinary garden soil.

For berries, remember to plant male and female plants—one male should be able to pollinate half a dozen female plants.

No regular pruning is required.

SOME POPULAR SPECIES	
H. rhamnoides *(Europe, temperate Asia)* Erect stems with sharp spines and narrow, silvery leaves. Small yellow flowers in mid spring on	previous year's shoots; the plant is grown mainly for the small bright orange berries clustered along the branches in autumn and winter. Medium.

Hippophae rhamnoides

Hoheria

SH(E)/AD/PS–FS

A small genus of evergreen and deciduous shrubs from New Zealand. The name is from a Maori word, *hoihere*. They need a favourable district and are not dependable in cold areas. Those described below are evergreen. They make good wall shrubs.

Hydrangea

SH–CL/AD/PS–FS

Hydrangeas are distinctive plants, sometimes tricky to grow well, but always impressive in flower. The flower heads of most species are flattened or dome-shaped, the inner fertile florets usually being surrounded by sterile florets round the edge. The most popular species is *H. macrophylla*, of which there are many varieties, divided into two groups: The Hortensias are the familiar mop-headed type with sterile florets forming a large globular head. The Lacecaps have a ring of coloured ray florets around a flat head of fertile flowers.

How to grow

A well-drained soil that will not dry out is ideal. A sheltered position is advisable. Mulch each spring.

How to grow

A well-drained soil and sheltered position are the main requirements. Regular pruning is not necessary, but cut out dead or damaged wood in mid spring.

Propagation

Seed can be sown in a propagator in spring. Pot up individually and overwinter in a cold frame, before planting out in late spring the following year.

Layers can be made in early autumn and severed from the parent plant about a year later.

SOME POPULAR SPECIES	
H. populnea *(New Zealand)* Much-branched shrub with evergreen or semi-evergreen, broadly ovate leaves. These are yellow-green edged deep green in the variety 'Variegata'. Saucer-shaped white flowers in late summer and autumn. Large.	**H. sexstylosa** *(H. populnea lanceolata) (New Zealand)* Upright habit. Shiny, leathery leaves, lance-shaped and toothed. White flowers 2.5 cm (1 in) across produced in mid and later summer. One of the hardiest species. Large.

Hoheria populnea 'Variegata'

Hydrangea macrophylla (Hortensia)

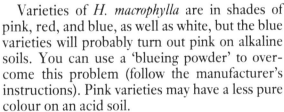

Above *Hydrangea macrophylla* (Lacecap)
Left *Hydrangea paniculata*

Varieties of *H. macrophylla* are in shades of pink, red, and blue, as well as white, but the blue varieties will probably turn out pink on alkaline soils. You can use a 'blueing powder' to overcome this problem (follow the manufacturer's instructions). Pink varieties may have a less pure colour on an acid soil.

Propagation

Seed is sometimes used to raise *H. petiolaris*. Sow under glass at 16°C (60°F). Germination usually takes from one to three months. Prick out seedlings into individual pots and overwinter under a cold frame. Pot on in early spring. Harden off and plant out in late spring or early summer.

Cuttings are the most reliable method. Select firm, semi-ripe, non-flowering green shoots, 10–13 cm (4–5 in) long, in early or mid summer. Root under a cold frame, but preferably in a propagator. Move into 9–10 cm (3½–4 in) pots when rooted, which usually takes about a month. Grow on and overwinter in a cold frame. Pot on in early spring. Harden off and plant out in late spring or early summer.

SOME POPULAR SPECIES

H. macrophylla *(H. hortensis) (China, Japan)* Rounded shrub with light green coarsely-toothed ovate leaves. Large flower heads—15 cm (6 in) or more across—from mid summer to early autumn. There are many varieties—either Hortensia or Lacecap (see above). Small to medium.

H. paniculata *(China, Japan)* Vigorous shrub with long upright but slightly arching shoots. Terminal almost pyramidal heads of white flowers in late summer and early autumn. 'Grandiflora' has massive flower heads and makes one of the showiest large shrubs. Large.

H. petiolaris *(Japan)*
Climber, clinging by aerial roots. Roundish-ovate dark green leaves turn yellow in autumn. Lax, flat heads of creamy-white flowers in early summer, up to 25 cm (10 in) across. Vigorous climber.

H. serrata *(H. macrophylla serrata) (Japan)* Smaller but hardier than *H. macrophylla*. Daintier flower heads 5–6.5 cm (2–2½ in) across from mid summer to late autumn. There are several varieties, in shades of blue, white, and pink. Small.

H. villosa *(China)* Rounded shrub with spreading habit and large lilac-blue flower heads in late summer. Best in partial shade. Medium.

Hypericum

SH(E, some)/O/SD–FS

Hypericum is an old Greek name used by Dioscorides. The genus is large and contains both deciduous and evergreen shrubs and sub-shrubs, many being partially evergreen in Britain. Some are excellent, even if vigorous, ground cover for difficult positions.

Hypericum x *moseranum* 'Tricolor'

How to grow
Cut back *H. calycinum* to within a few inches of the ground every second year. Other species need only have the previous year's shoots cut back to within a couple of buds of the old wood in early spring each year.

Propagation
Seed can be used for the species, although cuttings are better, and division is more sensible for *H. calycinum*.

Sow under glass (13°C/55°F) in spring, barely covering the seeds. Seedlings usually show in one to three months, when they are best pricked out into individual 8 cm (3 in) pots. Grow on and overwinter under a cold frame, potting on in early or mid spring. Harden off and stand outdoors for the summer. They should be ready for planting in autumn.

Cuttings should be 5–10 cm (2–4 in) long, from soft or semi-ripe wood, taken from late spring to mid summer. Root under a cold frame, and pot up singly into 8 cm (3 in) pots as soon as well rooted. Overwinter under the cold frame, harden off and plant out in late spring.

Divide *H. calycinum* in mid autumn or mid spring. Replant immediately or pot up and grow on for a year, then plant out.

SOME POPULAR SPECIES	
H. x moseranum *(Garden origin)* Low-spreading semi-evergreen with arching shoots. Clusters of yellow flowers, 6.5 cm (2½ in) wide, from mid summer to mid autumn. Excellent ground cover. 'Tricolor' has attractive variegation. Dwarf to small.	**H. patulum** *(China, Japan, Himalayas)* Dwarf deciduous or semi-evergreen with bright golden-yellow flowers up to 5 cm (2 in) across from mid summer to early autumn. 'Hidcote' (one of the finest hypericums) is sometimes listed as a variety of this species.

Ilex

Holly

TR–SH(E)/O/SD–FS

A genus of about 300 species of evergreen and deciduous trees and shrubs.

Male and female flowers usually appear on separate plants, and you will need to consider this if you want berries. Some varieties are male clone, some female. If you want only one plant, try 'J. C. van Tol'—it should produce berries without a neighbouring male.

Hollies will make trees or shrubs depending on how you train them.

How to grow
Variegated forms are best in full sun, but all-green hollies will do well even in shade.

Ilex x *altaclarensis* 'Lawsoniana'

Hollies transplant badly, so buy young container-grown plants and water thoroughly until established.

No routine pruning is necessary. Shaped trees or shrubs are clipped in mid or late summer. Remove any plain green shoots on variegated varieties as soon as they are noticed.

Propagation

Hollies are slow from seed, it is better to use cuttings.

Take 8 cm (3 in) semi-ripe heel cuttings in mid or late summer. Root under a cold frame, and move into small pots in mid spring. Harden off and stand outdoors for one or two years, potting on as necessary. Alternatively grow the cuttings on in a nursery bed instead of in pots.

Layers are easy. Peg down suitable shoots one or two years old in autumn, after slicing half way through the stem from the underside. They should be ready to sever and replant in two years.

Budding and grafting are used to grow variegated varieties as standards on *I. aquifolium* rootstocks, but this job is better left to a commercial grower.

SOME POPULAR SPECIES	
I. x altaclarensis *(Garden origin)* Green spiny leaves. Varieties include 'Golden King' (almost spineless leaves edged gold; female) and 'Lawsoniana' (yellow splashes). **I. aquifolium** *(Europe)*	Densely-leaved shrub or tree with pyramidal outline. Varieties include 'Argentea Marginata' (leaves edged white; male or female), 'J. C. van Tol' (almost spineless leaves; regular crop of fruit).

Jasminum

Jasmine

SH–CL/O/PS–FS

Most of the 300 or so jasmines are tender, but there is a handful of very important hardy or fairly hardy shrubs. Most of them are summer-flowering climbers, but *J. nudiflorum* is a scandent winter-flowering shrub.

How to grow

Jasmines will thrive in any well-drained soil, but apart from *J. nudiflorum* and *J. officinale* they will need a warm, sheltered, sunny position. *J. nudiflorum* will grow well in some shade, and even against a sunless wall, but avoid a position exposed to the force of cold winter winds.

Prune *J. nudiflorum* by cutting back sideshoots that have flowered, together with any very old or weak wood. Tie in new growths. The other species will need no regular pruning.

Propagation

Take semi-ripe cuttings in mid or late summer and root in a cold frame. When rooted (usually one to two months), pot up and overwinter in a frost-free greenhouse. Harden off and stand the pots outdoors in late spring. Plant in autumn.

Layers are easy. Peg down young growths in early autumn. Twist or kink the stems first. The young plants should be ready to lift in a year.

SOME POPULAR SPECIES	
J. nudiflorum *(China)* (Winter jasmine) Sprawling, lax, wall shrub. Bright yellow flowers on leafless stems from early winter to mid spring. Small to medium. **J. officinale** *(Northern Persia, India, China)*	Climber with pinnate leaves and fragrant white flowers in loose terminal clusters from early summer to mid autumn. 'Affine' has slightly larger flowers, tinged pink on the outside. Moderately vigorous climber.

Jasminum officinale

59

Kalmia

Calico bush, mountain laurel

SH(E)/AD/PS–FS

A genus of eight summer-flowering lime-hating shrubs, named in honour of Peter Kalm (1715–79), a Finnish pupil of the Swedish botanist Linnaeus, who travelled in North America.

The exquisitely shaped flowers are reminiscent of sugar icing.

How to grow

Kalmias need a moist, lime-free, preferably peaty, soil. Add plenty of peat when planting. They do best in partial shade and are good subjects for light woodland.

No regular pruning is necessary, but it is worth removing faded flower heads.

Propagation

Seed is not generally used, as cuttings and layers are easier.

Take cuttings of semi-ripe wood in mid or late summer and root in a cold frame. They should have rooted by spring, when they can be potted up. Harden off and stand outdoors for the summer, but give frame protection for their second winter. Pot on or plant out in spring and grow on for another year or two before planting out.

Layers are easy. Peg down suitable stems in early autumn. They should be ready for lifting a year later.

SOME POPULAR SPECIES	
K. latifolia *(Eastern North America)* Dome-shaped shrub. Glossy green evergreen leaves. Clusters of bright pink flowers in early	summer. The variety 'Clementine Churchill' has red flowers and 'Brilliant' has deep pink flowers that are crimson in bud. Medium.

Kalmia latifolia 'Brilliant'

Kerria

Jew's mallow

SH/O/PS–FS

This genus, named in honour of William Kerr (who collected in China for Kew), has only one species. It is the double-flowered form of this that is the most commonly grown.

How to grow

Any ordinary soil will suit the kerria, but avoid a very windswept site, and in cold areas plant against a wall or fence.

After flowering, prune all the shoots that have flowered back to strong new growth and thin out if necessary.

Propagation

Seed is little used as it is much easier to divide a clump or take cuttings.

Cuttings of semi-ripe wood are taken in mid or late summer and rooted in a cold frame. In about a month they should have rooted and be ready for potting. Overwinter in the cold frame. Harden off in spring, stand outdoors for the summer, plant in autumn.

Divide suckering clumps in mid autumn or early spring. Plant immediately.

SOME POPULAR SPECIES	
K. japonica *(China)* Upright, slightly arching green stems with bright green leaves. Single yellow	flowers in mid and late spring (double in 'Pleniflora'). 'Variegata' has white-edged leaves.

Kerria japonica 'Pleniflora'

Koelreuteria

Golden rain tree, pride of India

TR/O/FS

A genus of eight species named in honour of Joseph G. Koelreuter (1733–1806), who was Professor of Natural History at Karlsruhe. Only one of them is in general cultivation. The tree will take many years before it settles down to give a good display of flowers—and the milder the district the more likely you are to be rewarded.

The yellow flowers are star shaped and carried in long terminal panicles in mid summer. They may be followed by bladder-like fruits.

How to grow

Any good garden soil will suit, but for prolific flowering it needs a sheltered, sunny position.

Propagation

Sow seed in mid or late winter, after pre-treating by pouring hot (not boiling) water over the seeds and leaving to soak for 12 hours. Stand the sown containers outdoors for a month, then place in a propagator or warm greenhouse. Overwinter the seedlings under a cold frame, then grow in a nursery bed or pots for two or three years.

Koelreuteria paniculata

SOME POPULAR SPECIES	
K. paniculata *(Northern China)* Pinnate leaves, often colouring in autumn. Large, lax terminal	clusters of yellow flowers in mid summer, sometimes followed by green, flushed red, bladder-like fruits. Large.

Kolkwitzia

Beauty bush

SH/O/FS

A genus, named in honour of R. Kolkwitz, a professor of botany at Berlin, possessing only one species. It is grown mainly for its flowers.

How to grow

An easy-to-grow shrub that seems to thrive in most soils, including very chalky ground.

To keep the shrub compact, cut back stems that have flowered, once the blooms fade. Remove some of the older flowering stems entirely, along with any weak ones.

Propagation

Seed is not often used. If you use seed, soak in tepid water for 24 hours before sowing in containers, and stand outdoors for a month or two before giving warmth.

Cuttings of semi-ripe wood are taken in mid summer. Make them 10–13 cm (4–5 in) long, preferably with a heel, and root them in a cold frame. Pot up when rooted (it usually takes a month). Overwinter in the frame, and pot on in spring before hardening off to stand outdoors for the summer. Plant out in autumn.

SOME POPULAR SPECIES	
K. amabilis *(Western China)* Graceful, upright, twiggy shrub. Foxglove-like pink and yellow flowers in late	spring and early summer. 'Pink Cloud' (pink) is a popular and very lovely variety. Medium to large, very adaptable.

Kolkwitzia amabilis

Laburnum

Golden chain tree

TR/O/PS–FS

A small but invaluable genus of deciduous flowering trees or shrubs. The most common species are a familiar sight in late spring and early summer when they produce their pendulous racemes of slightly fragrant yellow flowers— sadly they are uninspiring for the rest of the year.

The seeds are very poisonous, which may deter you from planting where there are small children. The variety 'Vossii' produces few seeds.

How to grow

Any reasonable soil will suit laburnums, but they

Laburnum x *watereri* 'Vossii'

will do best on well-drained ground. Although hardy, it is best to avoid a very exposed position.

Propagation

Seed of the species can be sown in a cold frame in mid autumn, after soaking for 12 hours. Sow seed singly in small pots, covering lightly. Place in a propagator or warm greenhouse in mid or late winter—seedlings should appear in about a month. Grow on under a cold frame. Harden off and plant out in a nursery bed in late spring. They will be ready for final planting in two or three years.

Hybrids and named varieties are grafted in early spring, using whip-and-tongue or side grafts, but this is not easy.

SOME POPULAR SPECIES	
L. alpinum *(Southern and Central Europe)* (Scotch laburnum) Short, sturdily-trunked tree with broad head. Deep green trifoliate leaves. Long, pendulous clusters of pea-type yellow flowers in late spring and early summer. Medium. **L. anagyroides** *(L. vulgare) (Southern and Central Europe)* Bushy, wide-spreading tree. Yellow flowers in	drooping racemes 15–25 cm (6–10 in) long, in late spring and early summer. Small to medium. **L. x watereri** *(Garden origin)* A hybrid between the previous two trees. Masses of 30 cm (1 ft) long racemes of yellow flowers in late spring. 'Vossii' (it may simply be sold as *L.* 'Vossii') has particularly long racemes and is very free-flowering. Small.

Laurus

Sweet bay, bay tree, bay laurel

SH(E)/O/PS–FS

A genus of two or three species of evergreen shrubs, only one of which is widely planted. Male and female flowers are produced on different plants, the female bearing 12 mm (½ in) long purple-black berries—though these are not an important feature. The sweet bay is a popular plant, often trained as a clipped specimen. A single plant will provide a good supply of bay leaves for the kitchen as well as making an attractive container plant.

How to grow

Any reasonable soil will grow a good plant (use a loam-based compost such as John Innes No 3 for containers). In an exposed position the leaves are likely to be scorched by frost or cold winds, and an exceptionally cold winter may be enough to kill the shrub.

Tub-grown specimens can be trimmed to shape a couple of times during the summer, shrubs grown normally should not require pruning.

Propagation

Seed is not often used as cuttings are easier.

Take semi-ripe heel cuttings in mid or late summer. Root in a cold frame, and pot up the

Laurus nobilis

rooted cuttings in mid spring, and harden off to spend the summer in the open. Overwinter in the frame again, then harden off and plant out in their permanent positions in late spring.

SOME POPULAR SPECIES	
L. nobilis *(Mediterranean region)* Densely-leaved shrub with pyramidal outline.	Aromatic dark green, glossy leaves. Small yellowish-green flowers in mid spring. Large.

Lavandula

Lavender

SH(E)/O/PS–FS

This genus of aromatic shrubs takes its name from the word *lava* (to wash), a reference to its use in the preparation of lavender-water. The flowers can be dried and used in sachets and *pot-pourris.*

Lavandula angustifolia 'Hidcote'

How to grow

Lavenders grow well on a wide range of soils, but do best on well-drained soil in a sunny position. They tolerate chalky soils well.

Remove dead flower stalks once the flowers fade, then in early or mid spring prune back hard without cutting into old wood. Unfortunately the plants tend to become leggy after perhaps half a dozen years, so it is as well to propagate some replacements.

Propagation

Seed can be sown in a cold frame in mid or late spring—but germination is erratic and may take a couple of months. Grow on in individual pots in the frame, removing the light (top) for the summer. Plant out the following spring.

Cuttings are the most popular method. Use semi-ripe shoots in mid summer, and root under a cold frame. They should root in about a month. Overwinter under the frame, then plant out.

SOME POPULAR SPECIES
L. angustifolia *(L. spica* in part, *L. officinalis) (Mediterranean regions)* (Old English lavender) Bushy shrub with narrow, silver-grey highly aromatic leaves and well-known spikes of greyish-blue flowers from mid summer to early autumn.   'Hidcote' is a compact form with narrow, grey-green leaves and violet   flowers in mid summer. It may be listed as 'Nana Atropurpurea', or just *L.* 'Hidcote'. 'Munstead' is another compact variety with lavender-blue flowers in mid summer. 'Vera' (usually referred to as Dutch lavender) and sometimes sold as *L. vera,* is a robust form with broader leaves, lavender-blue flowers. Small.

Leycesteria

Flowering nutmeg, pheasant berry

SH/O/SD–FS

A small group of unusual shrubs with erect bamboo-like growth, named in honour of Wm Leycester, a nineteenth-century Chief Justice of Bengal and a keen amateur gardener. Only one species is widely grown. It is often planted in pheasant coverts (most birds like the purple berries), but it is also an excellent garden shrub, the green stems being an attractive feature in winter.

How to grow

Any reasonable soil will do, but well-drained ground is best. Shade is tolerated, but the shrub will flower more freely in full sun.

Leycesteria formosa

Prune back shoots that have flowered, along with any damaged shoots to a few inches above ground level in early spring.

Propagation

Seed, less popular than cuttings, can be sown in mid or late winter and stratified for a month or two before being brought into warmth (about 16°C/60°F). Prick out (usually after about a month) into individual pots, harden off and stand outdoors for the summer. Grow on for another 15 months or so before planting out.

Hardwood cuttings, about 23 cm (9 in) long, are taken in mid or late autumn. Root in a cold frame, and remove the lights (tops) in late spring. Plant out in their final positions in autumn.

SOME POPULAR SPECIES	
L. formosa *(Himalayas)* Erect, bamboo-like glaucous green stems and heart-shaped mid green leaves. Small white	flowers surrounded by claret-coloured bracts, in dense terminal panicles. Purplish-black, round, fruits in autumn. Medium.

Ligustrum

Privet

SH(E)/O/SD−FS

A genus of about 45 deciduous and evergreen shrubs and small trees. Most gardeners think of the common green privet hedges when this genus is mentioned, but there are some highly desirable species and varieties that make fine garden shrubs.

How to grow

These are very accommodating plants and should do well in any soil (including chalk), and in sun or shade. Little or no trimming will be necessary if grown as an ordinary shrub.

Propagation

Seed is a useful method for common privet. Sow in containers in a cold frame in mid autumn. Prick out into a nursery bed in the open in autumn or spring, 12 to 18 months after sowing. Grow on for a further 12 to 18 months, when the young plants should be ready to plant out in the desired position.

Cuttings are the main method of propagation. Prepare semi-ripe heel cuttings in mid summer and root in a cold frame. Harden off and plant out in a nursery bed in mid spring, to grow on for another six or twelve months.

SOME POPULAR SPECIES	
L. japonicum *(Korea, Japan)* (Japanese privet) Bushy shrub with dense, compact habit. Shiny, dark green oval leaves. Heads of white flowers in late summer. Makes a good hedge or screen. Medium. **L. lucidum** *(China, Japan, Korea)* Erect shrub or small tree. Narrowly oval glossy leaves. Erect terminal heads of white flowers in late summer and early autumn. Small to medium tree. **L. ovalifolium** *(Japan)* (Oval-leaved privet) Widely-grown hedging plant with dark green	leaves. The golden form 'Aureum' (golden privet) is the one usually grown as a decorative shrub. Evergreen (semi-evergreen in a severe winter). Medium or large shrub. **L. vulgare** *(Europe)* (Common privet) Widely-grown hedging plant, evergreen in mild areas, deciduous or semi-evergreen in very cold winters. Green, narrowly oval to lance-shaped leaves. Clusters of dull white flowers in early and mid summer. Small black berries, soon eaten by birds. Medium to large shrub.

Ligustrum ovalifolium 'Aureum'

Liquidambar

Sweet gum

TR/O/PS–FS

A small genus of deciduous trees with maple-like leaves that usually develop rich autumn tints. The name of the genus comes from the words *liquidus* (liquid) and *ambar* (amber), referring to a

fragrant resin prepared from one of the species.

How to grow

Will do well on any fertile soil, but will not be successful on thin, chalky ground.

Propagation

Seed is slow and not often used—but if you do not have a small tree to layer it may be the only option. Sow under a cold frame in autumn, but be prepared to wait—germination is erratic and can take up to two years, and the plants will have to be grown on for a further four years or so before they are large enough to be planted out.

Layers are pegged down in early spring, first removing a sliver of bark on the underside of the stem. Lift rooted layers in autumn (it usually takes two years). Grow on in a nursery bed.

SOME POPULAR SPECIES	
L. styraciflua *(North America)* Upright tree with slender branches forming a narrow pyramidal head. Dark green maple-shaped	leaves becoming orange and scarlet in autumn. There are variegated varieties such as 'Aurea' (mottled yellow). Medium to large.

Liquidambar styraciflua

Lonicera (shrubby)

Shrubby honeysuckle

SH(E, some)/O/PS–FS

A genus of about 200 species of shrubs, some of them climbers, named in honour of Adam Lonicer, a German botanist. Most of them are very fragrant.

How to grow

Ordinary but well-drained soil is adequate for the shrubby species. Regular pruning is not necessary, but it is worth cutting back shoots that have flowered once the blooms have faded. Foliage types can be clipped to shape as required.

Lonicera nitida 'Baggesen's Gold'

Propagation

Seed is slow and the results variable. Cuttings are easy and more dependable.

Take cuttings of semi-ripe wood in mid summer and root under a cold frame. Pot up individually in spring or plant in a nursery bed. In either case grow on for six months to a year then plant out.

Layers are pegged down in autumn after first removing a sliver of bark on the underside of the stem. Use young shoots. They should be ready to lift in 12–18 months.

SOME POPULAR SPECIES			
L. fragrantissima *(China)* Upright, semi-evergreen shrub with ovate, stiff and slightly leathery leaves. Fragrant creamy-white flowers about 18 mm (¾ in) long in winter. Small to medium.	**L. nitida** *(China)* Densely-leaved evergreen grown for its small, oval, glossy green leaves. Often clipped in to a hedge. 'Baggesen's Gold' has yellow leaves becoming pale greenish-yellow in autumn. Small to medium.	**L. pileata** *(China)* Low, spreading evergreen or partially evergreen shrub, the branches often horizontal, making it an excellent ground cover. The inconspicuous flowers are followed by violet berries. Small.	**L. standishii** *(China)* Deciduous or semi-evergreen winter flowers, similar to *L. fragrantissima*, differing in botanical details. Red berries in summer. *L. s. lancifolia* is a narrow-leaved form. Medium

Lonicera (climbing)

CL(E, some)/O/PS–FS

The shrubby honeysuckles were included in the previous entry. The climbers are useful for clambering over fences or arches, or scrambling through trees.

Lonicera x *americana*

How to grow

Like clematis, these climbers also like a cool, shady root-run with their heads in the sun. Incorporate plenty of manure or garden compost when planting, and mulch each spring.

Once flowering is over, remove surplus shoots to keep the plant within bounds, taking out a few old stems too.

Propagation

Layers or cuttings are the easiest methods of propagation—see shrubby species.

SOME POPULAR SPECIES	
L. x americana *(Garden origin)* Free-flowering deciduous climber. Fragrant white flowers, passing to yellow, tinged purple outside, early and mid summer. Vigorous. **L. periclymenum** *(Great Britain, Asia Minor, the Caucasus, Western Asia)* Deciduous climber with ovate to obovate dark green leaves. Terminal clusters of pale yellow flowers flushed purple-red outside, mid to late summer. 'Belgica' (early Dutch honeysuckle) is	bushier and flowers in early summer. 'Serotina' (late Dutch honeysuckle) flowers in mid autumn. Vigorous. **L. japonica** *(Japan, Korea, China)* (Japanese honeysuckle) An evergreen climber with light green ovate to oblong leaves on slender twining stems. White or pale yellow fragrant flowers from early summer to autumn. 'Aureoreticulata' has yellow veins. Severe weather may cause its leaves to fall. Moderately vigorous.

Magnolia

SH – TR(E, some)/O/PS – FS

A genus of 80 species of hardy evergreen and deciduous flowering trees and shrubs, most of them with magnificent flowers. The genus was named in honour of Pierre Magnol, a professor of botany and medicine at Montpellier in the sixteenth century.

How to grow

Most magnolias do poorly on limy soil, and do best on good, well-drained loam. Mulch young plants with peat or garden compost while young and becoming established. Plant in mid spring and choose a position sheltered from cold winds.

No routine pruning is required.

Magnolia x *soulangiana* 'Lennei'

Propagation

Seed is best sown in containers under a cold frame in mid autumn, using a peaty compost and barely covering the seeds. They should be ready for pricking out in 9 to 18 months. The following spring plant outdoors and grow on for three or four years before planting out.

Cuttings are useful and not too difficult. Use semi-ripe wood taken with a heel, and root under glass, ideally in a propagator, in mid summer. Pot up singly when rooted (in about six weeks). Pot on in the following spring then treat as seed-raised plants. Use a lime-free compost.

Layering is the easiest method if practical. Peg suitable shoots down in spring. The layers should be ready for planting out in two to two-and-a-half years.

Grafting onto pencil-thick seed-raised rootstocks in early spring is a feasible but specialised method.

SOME POPULAR SPECIES

M. grandiflora *(USA)*
Evergreen tree or large shrub with oval to oblong dark green, glossy leaves. Huge, fragrant, globular creamy-white flowers from mid summer to early autumn. Usually trained against a wall. Large shrub or small tree.

M. liliiflora *(Japan)*
Deciduous shrub with open habit. Oval to pear-shaped dark green leaves, and chalice-shaped purple flowers, white within, from mid spring to early summer. 'Nigra' has slightly larger deep purple flowers, creamy-white flushed purple inside. Medium shrub.

M. x loebneri *(Garden origin)*
Deciduous tree or shrub. Reddish-purple chalice-shaped flowers 7.5 – 10 cm (3 – 4 in) wide, in early and mid spring. Flowers young, and does well even on chalk. Small tree or medium shrub.

M. sieboldii *(M. parviflora) (Japan, Korea)*
Slender branches with dark green lance-shaped leaves, downy beneath. Bowl-shaped, pendent, pure-white flowers, cup-shaped at first. Deciduous. Small tree or large shrub.

M. x soulangiana *(Garden origin)*
Wide-spreading tree. Chalice-shaped white flowers, about 15 cm (6 in) wide, rose-purple at their bases, mid spring to early summer. There are several varieties, including 'Lennei' (rose-purple, white inside), and 'Alexandria' (free-flowering; large white blooms flushed purple at the base). Small tree.

M. stellata *(Japan)*
Compact deciduous shrub or small tree. White, star-like flowers, 7.5 – 10 cm (3 – 4 in) wide, in early and mid spring. Small tree or large shrub.

Mahonia

SH(E)/O/SD–FS

Top *Mahonia aquifolium*
Bottom *M. lomariifolia*

Hardy evergreen shrubs grown for their attractive foliage and yellow flowers. The genus honours Bernard M'Mahon (1775–1816), an American horticulturist.

How to grow

Mahonias grow well on most soils, including chalk, in sun or shade. Pruning is not normally necessary although unwanted shoots can be cut back in mid spring.

Propagation

Mahonias are easy to grow from seed, but it is best prechilled for three weeks before sowing in late winter. Overwinter under a frame. Pot on in spring before hardening off and growing outdoors for a year or two before final planting.

Cuttings are fairly easy. Use semi-ripe tips and root under glass, ideally in a propagator, in mid summer. Pot up singly when rooted, which takes about six weeks, and overwinter in a cold frame. Pot on in spring and treat as for seed-raised plants.

Suckers can be detached in mid autumn or mid spring, and grown on in a nursery bed for two or three years. This is an easy method of propagation if you choose sturdy suckers.

SOME POPULAR SPECIES

M. aquifolium *(Western North America)*
(Oregon grape, holly-leafed berberis) Suckering shrub with leathery, dark green, glossy leaves formed of five to nine spine-toothed leaflets. Dense heads of fragrant rich yellow flowers in early and mid spring. Blue-black berries. Small.

M. bealei *(China)*
Upright shrub with grey-green leaves formed of broad, leathery leaflets. Erect terminal heads of lemon-yellow flowers in winter. Medium.

M. japonica *(China)*
Stiff, sturdy, upright shrub with unbranched stems bearing dark green, glossy leaves. Drooping racemes of fragrant, lily-of-the-valley scented, flowers from mid winter to early spring. Medium.

M. lomariifolia *(China, Taiwan)*
Erect-stemmed shrub with dark green leaves 45–60 cm (1½–2 ft) long, bearing up to 20 pairs of holly-like spine-tipped leaflets. Erect racemes of fragrant deep yellow flowers from mid winter to early spring. Not hardy in cold areas. Large.

M. pinnata *(South West USA)*
Dull greyish-green leaves formed of up to 13 spiny leaflets. Erect clusters of slightly fragrant rich yellow flowers in early and mid spring. Medium.

Malus

Flowering crab/ornamental crab

TR–SH/O/PS–FS

A genus of 35 species of flowering and fruiting deciduous trees or shrubs, including the apple. Most of the species below, although having small and not unattractive fruit, are grown primarily for their flowers and sometimes foliage. Although grown as ornamentals, their fruits can be used for making preserves.

How to grow

Crab apples thrive in any ordinary well-drained soil, but will do better if it is enriched with manure or compost when you plant, and mulched each spring for the first few years. No regular pruning is required, but straggly or damaged shoots can be removed in winter.

Propagation

Seed-raised plants are slow to flower—expect to wait ten years or so—and in any case seed is only suitable for the species. Budding and grafting are possibilities, but you will probably find it much more worthwhile to buy plants.

Budding is easier than grafting. In mid or late autumn bud the selected variety on to an apple rootstock, using the 'T' or chip-bud method. Grow on the young trees for three or four years before planting permanently.

Grafting is perhaps best left to commercial growers.

SOME POPULAR SPECIES	
M. floribunda *(Japan)* Round-headed tree, densely branched. Crimson flower buds opening to white or blushed pink, in late spring, followed by small red and yellow fruits. Medium tree. **M. hupehensis** *(China, Japan)* Stiff outline with ascending branches. Fragrant white flowers, soft pink in bud, turning white, in late spring and early summer. Yellow fruits tinged red. Medium	to large tree. **M. sargentii** *(Japan)* Bushy shrub with masses of white flowers, tinged pink, in late spring. Small, bright red fruits in autumn. Medium shrub. **M. tschonoskii** *(Japan)* Erect, conical habit. Grown mainly for its splendid autumn colour. White flowers tinged pink, in late spring, followed by small dull red fruits flushed yellow (these are not freely produced and are not very attractive). Medium tree.

Malus floribunda

Nyssa

Tupelo

TR/M/PS–FS

A genus of ten hardy deciduous trees of which only one is widely grown. It is an undistinguished tree for most of the year but the leaves take on brilliant autumn colouring.

How to grow

In nature the nyssas prefer swampy places, but they adapt well to an ordinary loam once established. Not suitable for limy soil.

Propagation

Seed is sown in lime-free compost in a cold frame in mid autumn. Germination commonly takes 8 to 20 months. Overwinter under the frame, harden off and plant out in a nursery bed to grow on for three to five years before planting.

Layers are easy but you need a tree with suitable low branching. Use new shoots, and peg them down in the autumn, after slicing half way through the stem from the underside. They should be ready for lifting in two years, but will have to be grown on for a year or two before planting.

SOME POPULAR SPECIES	
N. sylvatica *(Eastern USA)* Pyramidal tree with shiny, mid green oval leaves that	turn brilliant shades of red, orange, and scarlet during late autumn. Medium.

Nyssa sylvatica

Olearia

Daisy bush

SH(E)/O/FS

A genus of about 100 species of Australasian wind-resistant and sun-loving hardy or slightly tender evergreens. The name is believed to refer to a resemblance to *Olea* (olive): some species have olive-like leaves.

How to grow

Olearias are best grown in mild or maritime areas

(they are excellent coastal shrubs), or by a sheltered wall inland. Any reasonable well-drained soil will do. Dead-head faded blooms with shears. Remove any dead branches, or those damaged by frost, in spring.

Propagation

Cuttings of semi-ripe wood taken in mid summer root easily. Root in a cold frame then pot up individually (they usually root in about a month). Overwinter under the frame. *O. macrodonta* and *O.* x *scilloniensis* are best given a little winter warmth in a greenhouse. Pot on in mid spring, harden off and stand outside for the summer. Plant in early autumn.

SOME POPULAR SPECIES	
O. x haastii *(New Zealand)* Rounded bush with crowded ovate, glossy leaves, white-felted beneath. Clusters of daisy-like white flowers cover the bush in mid and late summer. Medium. **O. macrodonta** *(New Zealand)* (New Zealand holly) Leathery, holly-like mid green leaves, white	beneath. Flat clustered heads of white daisy-like flowers in early and mid summer. Medium to large. **O. x scilloniensis** *(Garden origin)* Bright grey-green narrowly oblong leaves. Free-flowering, covered with white daisy-like flowers from late spring and early summer. Small.

Olearia x *haastii*

Osmanthus

SH(E)/O/PS–FS

A genus of evergreen flowering shrubs. Only the species below are hardy, and these need protection from very cold winds. The name of the genus is derived from the words *osme* (fragrance) and *anthos* (flower), an allusion to the fragrant flowers.

How to grow
Needs a well-drained soil and a position sheltered from cold winds. No routine pruning is necessary.

Propagation
Cuttings of semi-ripe wood are taken with a heel in mid or late summer. Use a rooting hormone, and root under glass, ideally in a propagator. Under suitable conditions they should have rooted in about a month. Overwinter in a cold frame and pot on in mid spring. Harden and grow on for a year or two outdoors before finally planting out.

Layers are easy if you have suitable shoots. Gently twist and kink stems two or three years old, and peg down in spring. Cover the wounded area with fine soil. Lift and plant the layers in spring once they have rooted, usually after about two years.

SOME POPULAR SPECIES	
O. delavayi *(China)* Slow-growing shrub with small, glossy, dark green leaves. Profusion of fragrant, pure white tubular flowers in mid spring. Medium.	**O. heterophyllus** *(O. ilicifolius)* *(Japan)* Bushy, rounded shrub. Glossy, dark green, holly-like leaves and clusters of fragrant white flowers in autumn. Medium.

Osmanthus delavayi

Paeonia

Peony

SH/O/PS–FS

A genus of 33 species of hardy herbaceous and shrubby perennials. The name was used by Theophrastus, and is said to be from Paeon, the physician who first used the plant medicinally.

How to grow
A moist but well-drained soil is ideal for peonies. Make sure the soil contains plenty of humus by working in garden compost or well-rotted manure before planting, and mulch each spring. Choose a sheltered site if possible. Cut out dead wood in spring, otherwise no pruning is required.

Propagation
Seed is a useful method. Sow in containers under a cold frame in early autumn. Prick out into 10 cm (4 in) pots when germinated in the spring. Overwinter under the frame, harden off, and plant out in a nursery bed. Grow on for three or four years before planting in their permanent position.

Cuttings are fickle, with a low success rate. It is better to use layers.

Layers are easy. Peg down suitable shoots in early spring after removing a sliver of bark on the underside of the stems. The layers are usually ready to be severed and replanted after about two years.

Tree peonies are sometimes root grafted on to herbaceous peonies in late summer, but this is difficult and best left to the professionals.

Paeonia delavayi

Parrotia

TR–SH/O/PS–FS

A genus with only one species (there used to be two, but one was transferred to the genus *Parrotiopsis*). It is named in honour of F. W. Parrot (1792–1841), a German naturalist and traveller. The plant is grown mainly for its autumn tints, but these can be very variable, not only from year to year but from garden to garden. It will make a large tree eventually, but is slow growing.

How to grow
Best in moist but well-drained soil and tolerates lime well. It does not require pruning and is very trouble free.

Propagation
Seed is sown in containers, under a cold frame in mid autumn, but germination is slow and spasmodic and often takes more than a year. Pot up singly as soon as the first pair of leaves are open. Overwinter under the frame. Pot on in mid spring and stand outdoors for the summer. Grow on for four or five years, potting on as necessary, until ready to plant out permanently.

Layering is the best method of propagating parrotias. Peg down shoots two or three years old in early or mid autumn, after slicing obliquely half way through the stems from the underside. Lift and pot up the layers when rooted in about two years, then grow on for four or five years.

Parrotia persica

Parthenocissus
CL/O/SD–FS

A genus of about ten species of self-clinging, deciduous climbers. In Britain there is often confusion between two of these—the Boston ivy, *P. tricuspidata*, often being called the Virginia creeper (which is really *P. quinquefolia*).

How to grow
Prepare a large planting hole near the base of the wall and add plenty of well-rotted manure or garden compost. Water the plants thoroughly for the first year, as the ground near a wall is often dry even after a shower of rain. Pinch out the growing tip to encourage branching from a low level. You may need to support with twiggy sticks until they become self-clinging. *P. henryana* need a sheltered position and its variegation is usually best in shade.

Propagation
Seed is sown in a cold frame in mid autumn, preferably after soaking in tepid water for 12 hours. Prick out singly when the seedlings show in spring, grow on and overwinter in the frame. Harden off then stand outside for the summer, potting on and giving support when necessary. Plant out permanently in autumn.

Cuttings of semi-ripe wood taken in late summer or early autumn root readily in a propagator. Pot up singly and overwinter in a cold frame. Harden off and treat as for seedlings.

Parthenocissus quinquefolia

Hardwood cuttings taken in late autumn will root without difficulty under a frame or in a sheltered spot outdoors.

Layers are easy, and should be ready for permanent planting within a year. Peg them down in mid or late autumn.

SOME POPULAR SPECIES	
P. henryana *(China)* (Chinese Virginia creeper) Self-clinging climber with dark green, lobed leaves, variegated white and pink along the main veins. In autumn the green parts turn red, accentuating the variegation. Vigorous. **P. quinquefolia** *(North America)* (Virginia creeper) Self-	clinging climber, with three or five lobed leaflets, brilliant crimson in autumn. Very vigorous. **P. tricuspidata** *(Japan, China)* (Boston ivy) Self-clinging climber with variable leaves but usually three-lobed in older plants. Rich scarlet autumn colouring. Very vigorous.

Pernettya
SH(E)/AD/SD–FS

A genus of 20 species, named in honour of Antoine Joseph Pernetty (1716–1801), who accompanied Bougainville on his voyage to the Falkland Islands and South America, chronicling the incidents of the trip.

Only one species is widely grown, this primarily for its colourful berries in shades of pink and red as well as white.

Pernettya mucronata

How to grow

Plant in lime-free soil (a peaty loam is ideal). Add peat when planting. Pernettyas will grow in shade but the plants will have a better habit and probably have more berries in a sunny position.
Plant in groups of three to five plants if you want to be sure of a good set of berries.

Although regular pruning is not required, old plants tend to become tall and straggly, in which case cutting back into the old wood in early spring should encourage new growth. The plants may also need trimming back in summer if the plants outgrow their allotted space.

Propagation

Seed is rather unpredictable in results. Cuttings are a better proposition and they are moderately easy to root.

Make heel cuttings 5 cm (2 in) long from firm, semi-ripe wood, and root in a cold frame or in a cold greenhouse in early autumn. Select cuttings from both male and female plants, to ensure berrying. Pot up singly in mid spring, harden off, and grow on in the open for a year before finally planting out. Use a lime-free compost for both seeds and cuttings.

SOME POPULAR SPECIES	
P. mucronata *(South America)* Low-growing, suckering, thicket-forming shrub with small, dark green leaves. White, nodding, heath-like flowers in late	spring and early summer. Dense clusters of round berries in autumn, in shades of pink, purple, red, and white. There are several good named varieties. Small.

Perovskia

Russian sage

SH/O/FS

A small genus of sub-shrubs and herbs, only one of which is in general cultivation. The genus honours V. A. Perovski, once Governor of the Russian province of Ouenberg. A plant as at home in an herbaceous border as among shrubs.

How to grow

Best on well-drained, light soil. Does not object to lime, and thrives in maritime areas. A sunny position is essential for good results.

In spring, cut the stems back to about 30 cm (1 ft) above soil level.

Propagation

Cuttings are easy. Make them 8 cm (3 in) long from non-flowering shoots, taken in mid summer. They will root readily in a cold frame. Harden off and plant out in spring.

SOME POPULAR SPECIES	
P. atriplicifolia *(Afghanistan to Tibet)* Upright semi-woody shrub. Lavender-blue	flower spikes in late summer and early autumn. 'Blue Spire' has larger flowers. Small.

Perovskia atriplicifolia 'Blue Spire'

Philadelphus

Mock orange

SH/O/PS–FS

A genus of popular summer-flowering, fragrant shrubs. The name of the genus is derived from an ancient Greek work meaning brotherly love.

How to grow

These shrubs will tolerate a wide range of soils and conditions. Thin out old wood after flowering, but bear in mind that the young shoots will carry next year's flowers.

Philadelphus 'Beauclerk'

Propagation

Cuttings root easily. Take cuttings of semi-ripe wood in mid or late summer and root them in a cold frame, where they can overwinter. Pot up into 13 cm (5 in) pots in mid spring. Harden off in mid summer and plunge outdoors in a sheltered spot until planting time in autumn.

Hardwood cuttings, taken in mid autumn, will root in sandy soil under a frame or in a warm, sheltered position outdoors. These will normally be ready to plant out a year later.

SOME POPULAR SPECIES

P. coronarius *(Northern and Central Italy, Austria, Rumania)*
Dense, bushy shrub with erect stems. Prominently veined green, oval leaves. Usually grown as either 'Aureus' (yellow leaves, slowly becoming greenish-yellow), or 'Variegatus' (creamy-white edges). Medium.
P. x lemoinei *(Garden origin)*
Round-topped deciduous shrub. Very fragrant, pure white flowers, about 2.5 cm (1 in) across, in early summer. Small to medium.

P. microphyllus *(South West USA)*
Rounded, dense, bushy shrub. Fragrant pure white flowers in early summer.
Garden hybrids
Some of the best philadelphus are hybrids of the various species. Typical are 'Beauclerk' (milky white with flushed cerise-pink around the base of the petals), 'Bouquet Blanc' (orange-scented, double flowers), 'Sybille' (single white), and 'Virginal' (very fragrant double and semi-double flowers).

Phormium

New Zealand flax

SH(E)/O/FS

A genus of only two species, although there are many varieties, which bring much variation in height and colouring. The name of the genus comes from the Greek word *phormos* (basket), presumably because baskets were made from the fibres. The fibres in the leaves have long been used in New Zealand in the manufacture of rope and twine. They are more or less hardy in all but very cold areas.

Phormium tenax

How to grow

A deeply-cultivated, moisture-retentive soil is required for phormiums. Plant in a sheltered position in full sun.

Protect the crowns with bracken, straw, or peat in winter unless you live in an area with mild winters. They are unlikely to be damaged except in very hard winters, but the precaution is probably worthwhile.

Propagation

Seed can be sown in warmth under glass in early spring, but division is easier, quicker, and essential if you want particular named varieties.

Lift the plants in mid spring and carefully split into rooted clumps with three or four leaves each. Replant immediately.

SOME POPULAR SPECIES	
P. cookianum *(P. colensoi) (New Zealand)* Sword-like pale green leaves, in the variety 'Tricolor' margined creamy-yellow with a thin red stripe at the edges. Yellowish flowers in mid and late summer. Small.	**P. tenax** *(New Zealand)* Larger than previous species, and with dull red flowers. There are varieties with coloured foliage, such as 'Purpureum' (bronze-purple) and 'Variegatum' (yellow stripes).

Photinia

SH(E)/O/FS

A genus of evergreen or deciduous trees or shrubs, but only one—a hybrid—is in general cultivation. This is proving hardy in sheltered positions.

It has many of the merits of a pieris (which needs an acid soil), but can be grown on neutral and chalky soils.

How to grow

Will grow in any good soil, and tolerates chalk, but avoid heavy clay.

Cut back new growth once it has lost its colour, to encourage further new leaves.

Propagation

Cuttings are quicker than seeds, and it is in any case worth propagating the best named varieties.

Take semi-ripe heel cuttings in mid or late summer. Root in a propagator if possible. Pot up singly when rooted and overwinter under a cold frame. Pot on in mid spring, harden off, and plunge outdoors in a warm position. Grown on for another year before finally planting out.

SOME POPULAR SPECIES	
P. x fraseri *(Garden origin)* Leathery, ovate to lance-shaped dark green, glossy leaves, coppery when	young. The variety 'Red Robin' has brilliant red young growths. Medium to large shrub.

Photinia 'Red Robin'

Pieris

Lily-of-the-valley shrub

SH(E)/AD/SD–FS

Evergreen shrubs or small trees requiring an acid soil. These multi-merit shrubs have attractive flowers but are sometimes grown for the new leaves that open bright red or orange on some species.

How to grow

A lime-free soil is essential for good results. A sheltered position in partial shade will bring out the best in them.

Incorporate plenty of peat when planting and mulch with peat each spring.

Regular pruning is unnecessary, but it is worth removing dead flower heads in mid spring, at the same time lightly trimming back any straggly shoots.

Propagation

Seed can be tricky and slow, so cuttings are normally used.

Prepare semi-ripe cuttings in late summer, and root in containers in a cold frame. Pot up individually into 8 cm (3 in) pots in late spring.

Pieris formosa forrestii

Harden off and plunge outdoors in light shade. Grow on for two years, potting on in spring as necessary. Plant into their permanent positions in autumn. Use a lime-free compost.

SOME POPULAR SPECIES	
P. floribunda *(South East USA)* Bushy, rounded, densely-leaved shrub with dark green, leathery leaves. Abundant clusters of pitcher-shaped white flowers in mid and late spring. Small to medium. **P. formosa** *(Eastern Himalayas, Yunnan)* Spreading shrub with	lance-shaped glossy, leathery leaves, coppery-red when young. Drooping clusters of white, pitcher-shaped flowers in mid and late spring. *P. f. forrestii* (called *P. forrestii* by some) has brilliant red young leaves. Medium to large.

Polygonum

Mile-a-minute vine, Russian vine

CL/O/PS–FS

This large genus of about 300 species contains hardy annuals, herbaceous perennials, and deciduous shrubby climbers. The name is derived from *polys* (many) and *gonu* (a small joint or knee joint), which may allude to the fact that some species have swollen nodes, though some argue that it refers to the many joints or knots in the roots.

How to grow

A tough plant that will grow well on most soils, in sun or partial shade.

Polygonum baldschuanicum

It is rampant, and it may be necessary to cut back encroaching stems in spring to keep the plant within bounds.

This vine must have some form of support such as a wall, fence or old tree stump.

Propagation
Cuttings of semi-ripe wood, taken in mid or late summer (preferably with a heel), provide the most reliable means of increase. Insert in containers under a cold frame and pot up when rooted—it usually takes about a month. Overwinter in the frame, and pot on in spring. Harden off and plunge outdoors in late spring to grow on and be ready to plant out the following spring.

Hardwood cuttings taken in mid or late autumn will often root outdoors in a sheltered, warm corner in well-drained sandy soil. Lift and plant out when dormant in late autumn, about a year later.

SOME POPULAR SPECIES	
P. baldschuanicum *(Bilderdykia baldschuanicum, Fallopia baldschuanica) (South East Russia)* Twining climber with ovate or heart-shaped pale	green leaves. Masses of foam-like white flowers, often tinged pink, are produced from mid summer to early autumn. Rapid growth. Very vigorous plant.

Potentilla
Shrubby cinquefoil
SH/O/PS–FS

Potentilla fruticosa 'Elizabeth'

This large genus contains hardy annuals, herbaceous perennials, and deciduous flowering shrubs and sub-shrubs. *Potentilla* is derived from the Latin *potens* (powerful), alluding to the alleged medicinal properties of some of these plants.

How to grow
Best in light, well-drained soil, in full sun (though they will tolerate partial shade).

Cut out weak shoots in early spring. After flowering, tip back the faded flowering shoots.

Propagation
Seed can only be suggested as a method of raising potentillas if you want to experiment, as the results are usually uncertain.

Cuttings are far more dependable. Take them with a heel, from semi-ripe wood in mid or late summer. Root them under a cold frame, and leave them there to overwinter. Pot up in spring, harden off and plunge outdoors for the summer. Plant out in their final positions after another 18 months.

SOME POPULAR SPECIES	
P. arbuscula *(Himalayas, Northern China)* Small, rounded, densely-leaved shrub with deeply lobed pale to mid green leaves. Yellow flowers from late spring to late summer. Dwarf. **P. fruticosa** *(Northern hemisphere)* Compact, rounded shrub with bright buttercup-yellow flowers from late spring to late autumn. There are many varieties, including 'Katherine	Dykes' (primrose-yellow), 'Mandschurica' (mat-forming type with silver-grey leaves and white flowers), 'Red Ace' (red), and 'Tangerine' (coppery-yellow). Small. **P. parvifolia** *(Central Asia)* Compact shrub with semi-erect habit. Small leaves, small golden-yellow flowers produced in early summer (and some through till autumn). Small.

Prunus (deciduous)

TR/O/FS

A large genus of deciduous or evergreen trees and shrubs. Some of the popular evergreen species are listed in the next entry. It is the deciduous flowering trees that are the most popular and the most spectacular. These include ornamental almonds, peaches, plums, and cherries. Although most are grown primarily for their flowers, some have very good autumn colour in tints of brown, red and crimson on the foliage.

How to grow

The majority of species thrive in ordinary well-drained soil, and do well on chalk soils.

Regular pruning should not be necessary. If you have to cut out damaged or unwanted shoots, do so in late summer to reduce the chance of certain diseases entering the wood.

Propagation

Seeds (stones) take years to reach maturity, and are only worth sowing for fun.

Budding in mid summer or grafting in early spring are the main commercial methods used, but neither is easy for the amateur.

Cuttings of semi-ripe wood, taken in mid summer and kept warm and humid (ideally in a propagator) will often root—but do not expect a high success rate. Pot up when rooted, overwinter under a cold frame, then harden off in spring and grow on in a nursery bed for three to five years.

Layers provide the easiest means of propagation if you have suitable branches that you can peg down. Layer in spring and lift when rooted (usually in about two years), and plant in their final positions or grow on in a nursery bed.

Prunus 'Kazan' *is one of the most popular of all flowering cherries*

SOME POPULAR SPECIES

P. avium *(Europe)*
(Gean, mazzard, wild cherry) Pyramidal tree. Crimson tint in autumn. White, cup-shaped, scented flowers in mid spring. Small, glossy reddish-purple fruits. 'Plena' is a double form. Large.
P. cerasifera *(Balkan Peninsula)*
(Cherry plum, myrobalan) Round-headed tree. Clouds of small white flowers in late winter and early spring. Usually grown in the purple-leaved forms 'Pissardii'

(sometimes called 'Atropurpurea'; dark red young leaves, becoming purple), and 'Nigra' (leaves and stems dull purple, flowers pink). Medium.
P. dulcis *(P. amygdalus, P. communis) (North Africa to Western Asia)*
(Almond) Erectly-branched but spreading later. Pink flowers, about 2.5–5 cm (1–2 in) wide, in early and mid spring on naked branches. Varieties include 'Alba' (white) and 'Roseoplena' (double pink). Medium.

P. persica *(China)*
(Peach) Bushy habit. Pale pink flowers, about 2.5 cm (1 in) across, in mid spring. There are varieties with pink, red, or white flowers, single or double. Medium.
P. sargentii *(Korea, Japan)*
(Sargent's cherry) Young leaves open bronze-red. Single pink flowers in early spring. One of the best trees for autumn colour. Medium.
P. serrula *(Western China)*
Glossy, reddish-brown mahogany-like bark that

peels in strips. Clusters of small white flowers in mid spring. Medium.
Japanese cherries
(Garden origin)
Some of the most popular flowering cherries are the so-called Japanese cherries. Two of the most popular are 'Amanogawa' (narrow, upright habit, semi-double pink flowers, mid and late spring), and 'Kanzan' (wide tree, double deep pink flowers in late spring). There are dozens of other good Japanese cherries. Medium to large.

Prunus (evergreen)

SH(E)/O/SD–FS

Deciduous species were included in the previous entry. The evergreens, though less spectacular in flower, are very tough and dependable.

How to grow
Any reasonable, well-drained soil will be adequate, provided it is not too dry. Chalk is well tolerated. Trim in spring if necessary.

Propagation
Seed is slow, but can be sown in containers under a cold frame in autumn then given 16°C (60°F) in a greenhouse in late winter. After pricking out, gradually harden off and return to the cold frame to overwinter. Pot on, harden off, and plunge the pots outdoors. Grow on for two or three years.

Heel cuttings of ripe wood taken in late summer or early autumn will root fairly readily in a cold frame. To ensure future berrying, take cuttings from both male and female plants. Pot up rooted cuttings in spring, harden off, and plunge outdoors for two or three years before planting.

SOME POPULAR SPECIES

P. laurocerasus *(Eastern Europe, Asia Minor)*
(Laurel, cherry laurel) Quick-growing, wide-spreading shrub with oblong, leathery, glossy mid green leaves. Terminal clusters of white flowers in mid spring. Varieties include 'Magnoliifolia' (very large leaves), and 'Otto Luyken' (dense, dome-shaped, low growth habit and narrow

leaves). Small (certain varieties) to large.
P. lusitanica *(Spain, Portugal)*
(Portugal laurel) Wide-spreading, bushy plant with glossy, dark green leaves. Scented, creamy-white flowers in slender tassels in early summer, followed by dark purple fruits. 'Variegata' has white-edged leaves. Large.

Prunus laurocerasus

Pyracantha

Firethorn

SH/O/PS–FS

Popular shrubs usually grown for their bright berries. *Pyracantha* is derived from *pyr* (fire) and *acanthos* (thorn), a theme reflected in the common name—firethorn.

How to grow
Will grow well on most soils, including chalky ground.

If grown as a wall shrub, a trellis or wires will be necessary for training.

No pruning is necessary on free-standing shrubs, but wall-trained plants should have surplus growth cut away between late spring and mid summer—but bear in mind that if you cut out shoots that have carried flowers you will be cutting out berries too. The plants are spiny, so wear gloves when pruning or training.

Propagation
Seed can be used, but cuttings are more popular, and must be used if you want to propagate hybrids or named varieties.

Take semi-ripe cuttings in mid or late summer, and root in a propagator if possible. Pot up as soon as rooted, and overwinter under a cold frame. Pot on in early spring, harden off, and plunge outdoors for the summer. Plant out in autumn.

Pyracantha rogersiana 'Flava'

SOME POPULAR SPECIES	
P. angustifolia *(Western China)* Dense, spreading, bushy shrub with rather rigid horizontal branches. Small leaves, cream flowers in early and mid summer. Orange-yellow berries. Medium. **P. atalantioides** *(China)* Fast-growing, erect shrub. Clusters of white flowers in early summer. Long-lasting scarlet berries. Good for a sunless wall. 'Aurea' has yellow berries. Large.	**P. coccinea** *(Southern Europe, Asia Minor)* Similar to above species. Bright red berries. 'Lalandei' is stronger-growing and more erect than the type. Large orange-red berries. Large. **P. rogersiana** *(China)* Erect habit, rather pyramidal when young. White flowers in early summer, reddish-orange berries later. There are varieties with yellow and with bright red berries. Medium.

Pyrus

Ornamental pear

TR/O/FS

Pyrus was the Latin name for a pear tree, but there are some ornamental pears too. They tolerate atmospheric pollution.

How to grow
Will grow on almost any soil, even chalk, and once established should be trouble-free. No regular pruning is necessary.

Pyrus salicifolia 'Pendula'

Propagation

Budding or grafting on to pear or quince root-stocks is necessary when propagating the best varieties. As this involves obtaining and planting out suitable rootstocks two or three years old, and obtaining the scion or budwood before you start, it is a job for commercial growers to tackle.

Seed is not normally used either—even for rootstocks.

This is one tree where it is best to resign yourself to buying a plant.

Rhododendron (Azalea type)

Azalea

SH(E, some)/AD/PS

Azaleas are a form of rhododendron—they were previously given the status of a separate genus on the basis of having five stamens instead of ten, but now this is not considered sufficient to separate the azaleas and rhododendrons. There are two groups of azalea: the evergreen or Japanese type are low and spreading, the deciduous are taller and have good autumn colour.

How to grow

An acid soil is ideal, but rhododendrons can also be grown satisfactorily on a neutral soil. Add plenty of peat when planting and mulch with peat each spring. Choose a sheltered site with partial shade. Regular pruning is unnecessary, but remove dead flower heads (they will snap off).

Propagation

Seeds are a tricky and slow way to raise azaleas. One of the other methods is likely to be more successful.

Cuttings require careful nursing. Take 5–10 cm (2–4 in) semi-ripe cuttings of small-leaved varieties in early or mid summer and root under a frame. Pot up singly in spring if rooted; grow on under the frame for another 18 months, then in spring pot on and harden off before plunging outdoors to grow on for another two or three years. Use an acid compost at all stages.

Layering is the easiest way to increase azaleas—see rhododendrons. Grafting is also possible, but more difficult for amateurs.

SOME POPULAR SPECIES

Deciduous azaleas (*Garden origin*)
Flowering is usually in late spring and early summer, heights normally 1.5–1.8 m (5–6 ft). Wide colour range, with generally trumpet-shaped flowers. Good autumn leaf colour. Ghent hybrids usually have fragrant, long-tubed and honeysuckle-like flowers; Knap Hill and Exbury hybrids are usually unscented; Mollis hybrids have scentless but vividly coloured flowers that appear before the leaves. There are other groups and many varieties.
Evergreen azaleas (*Garden origin*)
Hardy, low-growing, spreading shrubs, flowering in mid and late spring—the foliage is often obscured by the mass of blooms. Again there are many groups, such as the Kurume azaleas, which have small flowers little more than 2.5 cm (1 in) across but are extremely free-flowering and colourful.

R. 'Irene Koster' (Occidentale hybrid)

Rhododendron

SH–TR(E)/AD/PS

There are over 500 rhododendron species (those included in the old *Azalea* genus are described in the previous entry). In addition there are thousands of hybrids and varieties.

The name *Rhododendron* is derived from the Greek *rhodon* (rose) and *dendron* (tree). In fact they range from trees down to dwarf alpine shrubs. The species described below are only representative examples.

How to grow
Treat as azaleas (page 83).

R. x 'Mrs P.D. Williams'

Propagation
Seed and cuttings are slow to reach flowering size (see azaleas for how to take cuttings), and layering is the best method.

Layering is easy and best carried out in spring or autumn, using shoots two or three years old. They will usually have rooted and be ready to lift in about two years. Lift in mid autumn or mid spring, and either plant in their final positions or pot up and grow on for between one and three years. Grafting is done commercially, but is a difficult method for amateurs.

SOME POPULAR SPECIES	
R. augustinii *(China)* Upright shrub with small dark green leaves and masses of funnel-shaped blue flowers in mid or late spring. Medium. **R. campanulatus** *(Himalayas)* Spreading shrub with leathery, dark green leaves, brown and woolly beneath. Bell-shaped flowers from pale rose to lavender in mid and late spring. Medium. **R. impeditum** *(China)* Grey-green leaves and light purplish-blue funnel-shaped flowers in spring. Dwarf. **R. wardii** *(Yunnan, South East Tibet)*	Compact habit. Clear yellow flower in loose trusses, sometimes with a crimson blotch at the base, in late spring. **Hardy hybrids** *(Garden origin)* There are hundreds of varieties, of which the following are examples. 'Blue Diamond' (lavender-blue flowers, mid spring; small), 'Bo-peep' (primrose-yellow, early spring; small to medium), 'Loder's White' (pink buds opening to white, late spring and early summer; large), 'Pink Pearl' (large pink flowers, late spring and early summer; large).

Rhus

Stag's horn sumach

TR–SH/O/FS

A genus of hardy deciduous shrubs and small trees, grown mainly for their attractive foliage, which colours well in autumn. *Rhus* comes from an ancient Greek name for the genus, used by Theophrastus. The smoke tree sometimes listed as *R. cotinus* will be found under its more correct name of *Cotinus coggygria* (page 33).

Rhus typhina 'Laciniata'

How to grow
These are undemanding plants and will grow in most soils. They do best in full sun.

Grown as a tree no regular pruning is necessary, but they can be grown as shrubby plants for their leaves alone by cutting the stems down to 30 cm (1 ft) above the ground each year in late winter.

Propagation
Cuttings of semi-ripe wood taken with a heel, in mid summer, will root without much difficulty, but a propagator is an advantage. Pot up singly and overwinter under a cold frame. Pot on in mid spring before hardening off and plunging outdoors for the summer. Young plants are normally ready for permanent planting by autumn.

Suckers provide the easiest method of increase. Carefully lift and detach healthy, well-rooted suckers in spring, and replant immediately.

Layers can be pegged down in spring, and should be ready to lift and plant out in a year or two.

SOME POPULAR SPECIES	
R. typhina *(Eastern North America)* Open, flat-topped tree, often branching from near soil level. Pinnate leaves	30–45 cm (1–1½ ft) long, with good autumn colour. 'Laciniata' has more deeply divided, fern-like leaves. Small tree.

Ribes

SH/O/PS–FS

A genus of 150 species of hardy and half-hardy deciduous and evergreen shrubs, including some with edible fruits, such as currants and gooseberries. The ornamental species are grown for their flowers.

How to grow
Ribes will grow well in any reasonable soil, in sun or partial shade. Prune after flowering, cutting back shoots that have flowered and taking old unproductive wood down to ground level.

Propagation
Seed is rarely used. Cuttings are more practical.

Root hardwood cuttings in sandy soil under a cold frame, or in a sheltered position outdoors, in mid or late autumn. Remove the frame lights (tops) in late spring. Lift the rooted cuttings in mid autumn and plant straight out into their permanent positions. Leave the less well rooted plants until the following spring or autumn.

Semi-ripe cuttings taken in mid or late summer will root readily under a cold frame and be ready for potting up in autumn. Overwinter under the frame, harden off in spring and plant out in a nursery bed. Plant out in mid autumn.

Ribes sanguineum 'Pulborough Scarlet'

SOME POPULAR SPECIES	
R. alpinum *(Northern and Central Europe)* (Alpine currant) Bushy, twiggy, shrub. Greenish-yellow flowers in erect bunches in mid spring. If male and female plants are present, round red berries follow in autumn. Tolerates shade well. Small. **R. odoratum** *(R. aureum) (Western North America)* (Buffalo currant) Rather lax shrub with crowded stems that arch outwards at their tips. Fragrant yellow flowers in mid spring. Purple-black fruits. Medium.	**R. sanguineum** *(Western North America)* (Flowering currant) Pendulous clusters of deep rose-red flowers in spring, followed by round, blue-black berries. There are several varieties, among them 'Brocklebankii' (yellow leaves, best out of full sun), 'Pulborough Scarlet' (deep red). Medium. **R. speciosum** *(California)* (Fuchsia-flowered gooseberry) Small clusters of slender, almost fuchsia-like bright red nodding flowers from mid spring to early summer. Medium.

Robinia
False acacia
TR/O/PS–FS

A genus of 20 hardy deciduous flowering trees and shrubs, native to the USA and Mexico. Jean Robin, herbalist to Henry IV of France, is the man who gave his name to this genus.

How to grow
Robinias will grow in almost any soil, in full sun or partial shade. No regular pruning is necessary.

Propagation
Seed is little used by amateurs, but you can try sowing it in spring in containers in a cold frame, soaking it for 12 hours first. Germination is likely to take six months, possibly over a year. The seedlings will need to be grown on outdoors for two or three years, potting on as necessary.

Hardwood cuttings taken in autumn will root readily under a cold frame. Uncover in late spring, pot up the rooted cuttings in the autumn, and plunge outdoors to grow on for a couple of years before planting out.

Suckers provide the easiest method of increase. Lift them with roots on in autumn, pot them up and grow on as for rooted cuttings. This method will not be suitable if the variety has been grafted.

Robinia pseudoacacia 'Frisia'

Grafting is used commercially for some named varieties, and *R. hispida* may be grafted on to *R. pseudoacacia*, but is difficult for the amateur.

SOME POPULAR SPECIES	
R. hispida *(South East USA)* (Rose acacia) Lax habit, but looks good trained against a wall. Dark green pinnate leaves. Drooping clusters of rose-pink pea-type flowers produced in late spring and early summer. Medium.	**R. pseudoacacia** *(Eastern USA)* (Common acacia, black locust, false acacia) Round-headed tree with light green pinnate leaves. Long, drooping clusters of creamy-white flowers in early summer. 'Frisia' has yellow foliage. Large.

Rosa
Shrub rose
SH/O/PS–FS

Mention a rose to most gardeners and they think of the modern large-flowered (hybrid tea) or cluster-flowered (floribunda) roses, but there are many species and their varieties and hybrids that can be grown as normal shrubs. There is a vast range from which to choose and the species included here represent only a small cross-section of some of the good shrub roses (both modern and old fashioned) available.

Rosa gallica 'Officinalis'

How to grow
These roses will grow on almost any soil, but do better if plenty of well-rotted manure or garden compost is added when planting. Species and 'shrub' roses require little pruning—just remove straggly growth each spring.

Propagation
Seed is little used, and most named varieties are unlikely to come true, but it is fun to try. Sow in containers in a cold frame in mid autumn. Pot up seedlings individually at the two-leaf stage, probably by summer. Overwinter in the frame. Harden off and plant out in a nursery bed in spring to grow on for another 18 months.

Hardwood cuttings are the normal method of propagation. Insert in sandy soil under a cold frame, or in a sheltered spot outdoors. Lift during autumn and plant out in their final positions when well rooted—about two years.

SOME POPULAR SPECIES			
R. centifolia *(Garden origin)* (Cabbage rose, Provence rose, Rose des Peintres) Erect, prickly stems with scented leaves. Large, fragrant double pink flowers in early and mid summer. 'Cristata' has crested sepals, 'Muscosa' has moss-like glandular bristles covering the	stems, flower stalks and calyces. Small. **R. damascena** *(Western Asia)* (Damask rose) Large, fragrant flowers in early and mid summer, bluish white to red, followed by red fruits. 'Versicolor' has loosely double white flowers blotched rose. Medium.	**R. gallica** *(Southern Europe, Western Asia)* (French rose) Suckering shrub with erect stems armed with small, slender prickles. Showy, purple-pink flowers in early summer, followed by round red hips. 'Officinalis' has semi-double rose-crimson, very fragrant flowers,	'Versicolor' has crimson flowers streaked white. Small. **R. moyesii** *(Western China)* Loose, open shrub. Blood-red flowers in early summer. Flagon-shaped crimson fruits. 'Geranium' is more compact, with bigger fruits. Large.

Rosmarinus

Rosemary

SH(E)/O/FS

A genus of about three evergreen shrubs. They are not totally hardy, but *R. officinalis* is likely to be injured only by very severe winters.

Rosemary tea was used to aid digestion and was said to ease neuralgic pain. The ancient Greeks claimed that it had properties that strengthened the memory—hence the saying 'Rosemary for remembrance'.

How to grow
Ordinary, well-drained soil and a sunny position will suit rosemary. It does well on chalky soil but not on clay. Trim the bush lightly with shears once flowering is over. A straggly plant can be cut back by half in mid spring.

Propagation
Sow seed in containers in early spring and maintain about 16°C (60°F). Pot seedlings singly and harden off gradually before plunging the pots outdoors for the summer. Give winter protection, then plant out in late spring.

Semi-ripe cuttings taken in mid summer are also easy. Insert under a cold frame and pot up singly when rooted. Overwinter in the frame, harden off, and plant out in late spring.

SOME POPULAR SPECIES	
R. officinalis *(Southern Europe, Asia Minor)* Erect and spreading shrub with narrow dark green leaves, white beneath, and mauve flowers in early and	mid spring, and intermittently through to late summer. 'Miss Jessop's Upright' has a more strongly erect habit. Medium.

Rosmarinus officinalis

Rubus

SH(E, some)/O/PS−FS

The size of this genus depends on the authority that you take, but it certainly runs into hundreds of species, most of them scrambling and usually prickly shrubs.

The species below highlight how diverse— and desirable—some of the relatives of the blackberry can be. They are useful for training up walls or fences or as ground cover.

How to grow

Will thrive in almost any soil, and most are best in partial shade.

Cut out a proportion of older, flowered, wood on flowering types in autumn, taking the shoots back to ground level. Prune *R. cockburnianus* in autumn by removing the shoots formed the previous year.

Propagation

Cuttings of semi-ripe wood, taken in late summer or early autumn, root readily under a cold frame. Insert singly in small pots and over-winter under the frame. Harden off the young plants and set out in a nursery bed in late spring. Grow on there and plant out in final positions in autumn.

SOME POPULAR SPECIES	
R. cockburnianus *(R. giraldianus) (China)* Deciduous shrub with upright stems covered with a white bloom, arching at the tips. Purple flowers in early summer, but not a feature. Grown mainly for decorative stems. Medium. **R. tricolor** *(Western China)* Evergreen or semi-evergreen carpeter. Long, trailing stems with red bristles. Glossy green	leaves, white beneath. White flowers in mid summer, sometimes followed by large red fruits. One of the quickest ground-covers for shade. Prostrate. **R. x tridel 'Benenden'** *(Garden origin)* Deciduous shrub with lobed leaves on tall, arching, spineless stems. Glistening white flowers 5 cm (2 in) across, produced in late spring. Quick growing. Medium.

Top *Rubus* x *tridel*
Bottom *Rubus cockburnianus*

Salix

TR – SH/M/PS – FS

A large genus of deciduous trees and shrubs, supplemented by many natural hybrids. They vary in size from dwarf shrubs to large trees. Some are grown for coloured shoots, and a few for their catkins (male and female catkins normally appear on separate plants). In most species it is the male catkins that are the showy ones.

How to grow

Salix do best in moist soil in a sunny position.

Regular pruning is not required, but those grown for a colourful crop of stems should be cut back to within a few inches of the ground annually in late winter.

Propagation

Hardwood cuttings taken from early autumn to early winter will root readily. Make them at least 20–35 cm (8–14 in) long and root in a sandy, well-drained soil under a cold frame or in a warm sheltered position outdoors. Remove the frame lights (tops) by late spring. Rooting usually takes from 12 to 18 months. Lift the rooted cutting to plant out in mid or late autumn or in early spring.

Layers are easy. Peg down stems one to three years old, in autumn or spring. They should have rooted and be ready for lifting from 12 to 18 months. Plant straight into their final positions.

Budding can be done in mid summer or grafting in early spring, and is usually done for weeping varieties. These are not easy methods.

SOME POPULAR SPECIES	
S. alba *(Europe, Northern Asia)* (White willow) Elegant tree with branches that droop at their tips. Narrowly lance-shaped leaves, silver beneath. Among the varieties are 'Chermesina' (scarlet willow) with brilliant orange-scarlet young stems in winter, and 'Vitellina' (golden willow), which has yellow young shoots. Large tree. **S. caprea** *(Europe, North West Asia)* (Goat willow, goat sallow) Low tree or shrub with bushy habit. Oval to lance-shaped grey-green, wrinkled and slightly downy leaves. The well-known 'pussy willow' catkins appear in early spring. 'Pendula'	(Kilmarnock willow) has stiffly pendulous branches. Small to medium tree. **S. x chrysocoma** *(S. alba* 'Tristis') *(Garden origin)* Weeping tree with long slender branches that hang down to soil level. Medium tree. **S. hastata** *(Europe, Northern Asia)* (Halberd-leaved willow) Hairy young shoots, purplish in second year. Oval to pear-shaped leaves. 'Wehrhahnii' has a spreading habit and silvery male catkins that turn yellow. Small shrub. **S. lanata** *(Europe, Asia)* (Woolly willow) Low, sturdy, spreading shrub. Silvery leaves and erect yellowish-grey woolly catkins. Small shrub.

Salix lanata is a small, slow-growing shrub that is suitable for the rock garden

Salvia

SH(E)/O/FS

A large genus of perhaps 700 species, containing annual, biennial, and perennial herbs, shrubs, and sub-shrubs. *Salvia* is an ancient Latin name used by Pliny, derived from *salveo* (to save or heal), alluding to the alleged medicinal properties of some species.

How to grow
Best in a light, well-drained soil in a sunny position.

May be damaged by severe frost, but soon recovers. In early spring cut back to within two or three buds of the old wood, otherwise the plants tend to become leggy with age.

Salvia officinalis 'Tricolor'

Propagation
Seed can be sown at 18–21°C (65–70°F) in early or mid spring. Prick out singly when large enough to handle—usually in six to eight weeks. Grow on a little cooler for a couple of weeks then move to a cold frame. Leave the lights (tops) off for the summer, but replace for the winter. Harden off and plant out in late spring. An easy method of propagation, but less popular than cuttings.

Cuttings of semi-ripe wood, taken from mid summer to early autumn, should root fairly easily and provide the best method of increase. Root in containers in a cold frame (or better still in a propagator). Pot up into individual pots when rooted (usually after five or six weeks), and over-winter under a cold frame. Harden off and plant permanently in late spring.

Layers root readily if firm young shoots are pegged down in early autumn. Young plants should be ready to lift and plant out in a year.

SOME POPULAR SPECIES	
S. officinalis *(Southern Europe)* (Common sage) Aromatic sub-shrub, the wrinkled leaves being used to flavour food. However, it is usually varieties with coloured leaves that are used as ornamental plants.	Popular ones are 'Icterina' (green and gold leaves), 'Purpurascens' (purple-leaved sage, stems and leaves soft purple), and 'Tricolor' (grey-green leaves splashed creamy-white and suffused purple and pink). Dwarf.

Sambucus

Elderberry

SH–TR/O/SD–FS

Hardy deciduous shrubs or small trees, grown as ornamentals mainly for their pinnate leaves, which are very attractive in some of the varieties if not the species. They tolerate almost all soils and situations.

How to grow
Almost any soil will suit them, and although they will tolerate shade they are best in full sun.

Sambucus racemosa 'Plumosa Aurea'

Yellow-leaved forms in particular need good light.

When growing for foliage effect, prune hard each spring, cutting the stems back to within a few inches of the ground.

Propagation

Seed germinates slowly and is little used as a method of increase.

Hardwood cuttings taken in late autumn root easily in sandy soil. Place in a cold frame or in a sheltered position outdoors. Remove frame lights (tops) in late spring. Lift the rooted cuttings in mid autumn and plant in their final positions.

Cuttings of semi-ripe wood root without difficulty if inserted under a cold frame in mid or late summer. Pot up individually and overwinter in the frame, then harden off in mid spring to plant outdoors in a nursery bed for the summer. Plant in final positions in mid autumn.

SOME POPULAR SPECIES	
S. nigra *(Europe)* Pinnate leaves. Yellowish or dull yellow flowers in flattened heads in early summer. More attractive are some of the varieties. These include 'Aurea' (yellow), 'Laciniata' (finely cut, fern-like leaves), and 'Purpurea' (flushed purple, especially young leaves). Large shrub or small tree.	**S. racemosa** *(Europe, Asia Minor, Northern China)* Mid green leaves formed of five oval, toothed leaflets. Clusters of yellowish-white flowers in mid and late spring, followed by scarlet berries. The most attractive form is 'Plumosa Aurea', which has deeply divided golden leaves and yellow flowers. Medium to large shrub.

Santolina

SH(E)/O/FS

A small genus of aromatic, hardy evergreen dwarf shrubs, or sub-shrubs, with feathery foliage. The name is derived from *Sanctum linum* (holy flax), an old name for *S. virens*.

How to grow

Any well-drained soil will suit them. Remove dead flower stems, and cut back hard in mid spring every third year to keep the plants compact and vigorous.

Propagation

Seed is sown in spring, ideally in a temperature of about 16–18°C (60–65°F) in early or mid spring. Grow on and harden off before plunging the pots outdoors in early summer. Give winter protection and plant out finally in late spring of the second year.

Semi-ripe cuttings root quickly and easily in containers under a cold frame if taken in mid or late summer. Make them 5–8 cm (2–3 in) long with a heel. Overwinter under cover and pot up singly in mid spring. Harden off gradually and plunge outdoors in late spring to grow on until autumn, when they can be planted out in their final positions.

SOME POPULAR SPECIES	
S. chamaecyparissus *(S. incana) (Southern France)* (Lavender cotton) Forms a dense mound of silvery-grey, woolly, finely dissected leaves. Bright lemon-yellow flowers in mid summer. *S. c. corsica* is a more compact variety. Dwarf. **S. neapolitana** *(Italy)* Similar to previous species, but looser growth	and more feathery foliage. Dwarf. **S. virens** *(S. viridis) (Origin uncertain, probably Southern Europe)* (Holy flax) Wide-spreading, mound-forming shrub with green stems and deep green thread-like leaves. Bright lemon-yellow flowers on slender stalks, mid summer. Dwarf.

Santolina chamaecyparissus

Senecio

SH(E)/O/FS

A very large genus containing about 3,000 species, ranging from tender succulents and half-hardy annuals to evergreen shrubs and trees. The name is an old one used by Pliny, from *senex* (an old man), referring to the white or grey hair-like seed pappus conspicuous on some species.

The naming of some of the popular species is unfortunately rather confused.

How to grow
Almost any well-drained soil will suit the hardy shrubby senecios, and chalk poses no problem.

Cut out straggly shoots in spring to keep the plants tidy. Cut off faded flowers.

Propagation
Cuttings of semi-ripe wood taken in mid or late summer are the usual method of increase. Root in containers under a cold frame. Pot up singly when rooted (it usually takes six to eight weeks), and overwinter under the frame. Pot on in mid spring before hardening off in late spring and plunging the pots outdoors. They should be ready for final planting in mid autumn.

Senecio greyi

SOME POPULAR SPECIES	
S. greyi *(New Zealand)* The plant sold under this name is usually *S. laxifolius* or a hybrid called 'Sunshine'. Mid green to ovate leaves densely covered with grey-white hairs. Terminal heads of clear yellow flowers in mid and late summer. Small. **S. laxifolius** *(New Zealand)* Hardier and more reliable than the previous species, but rather lax growth otherwise similar. The plant sold is often a hybrid—'Sunshine'.	**S. monroi** *(New Zealand)* Densely-foliaged, dome-shaped shrub with wavy-edged green leaves grey-white beneath. Heads of bright yellow flowers in mid summer. Small. **S. 'Sunshine'** *(Garden origin)* Until comparatively recently this plant was sold variously as *S. greyi* and *S. laxifolius*. You may still find it under those names, but the new name is becoming more widely used. Whatever the name, they are all good shrubs.

Skimmia

SH(E)/AD/SD–FS

Hardy, evergreen, bushy shrubs with male and female flowers usually on separate plants, and both sexes are needed to produce the bright red berries, which are one of the main attractions. If you want a bisexual species, grow *S. reevesiana*.

How to grow
Most species will grow well in any well-drained soil, but some, especially *S. reevesiana*, do not do well on chalky soil, and all do best on acid soil. Although they tolerate shade, best results are obtained in partial shade or full sun.

Skimmia japonica

Propagation

Seed is slow, and cuttings or layers are much easier.

Heel cuttings of semi-ripe wood root with little difficulty if taken in mid or late summer. Select from both male and female shrubs to ensure berrying. Root in a cold frame, and pot up the young plants in mid spring. Grow on for another year, protecting for the winter. Pot on in mid spring, then harden off to plunge outdoors again to grow on for another year.

Layers pegged down in autumn or spring are an easy alternative. They should be ready to lift in about two years. Lift in spring, pot up, and plunge outdoors to grow them on for a year or two before finally planting in autumn.

SOME POPULAR SPECIES	
S. japonica *(Japan)* Low, compact shrub, often forming a spreading clump. Leathery, ovate to lance-shaped leaves. Clusters of star-shaped creamy-white flowers in mid and late spring. Bright red berries in autumn if both sexes present. 'Foremanii', a vigorous female variety, will produce large bunches of red berries. 'Rubella' is a male variety with red buds	throughout winter, opening as white flowers in early spring. Small. **S. reevesiana** *(China)* Male and female flowers on the same plant, so a good choice if you want only one plant. Compact, crowded with dome-shaped heads of creamy-white flowers in late spring. Crimson berries by early autumn, lasting into winter. Small.

Sorbus

TR/O/PS–FS

Hardy deciduous trees and shrubs. *Sorbus* comes from the Latin *sorbum* a name used by Pliny and Cato for the fruits of *S. domesticum.*

How to grow

Adaptable trees, thriving in most soils, in sun or partial shade.

Propagation

Seed is slow, but can be sown in containers placed under a cold frame in mid autumn. Germination is erratic and may take a year or more. Prick out seedlings when large enough. Overwinter in the frame, then harden off and plant out year-old seedlings into a nursery bed in spring. Grow on for three to five years before final planting.

Budding in mid or late summer and grafting in early spring are methods used commercially to propagate varieties, but these are difficult for the amateur. After budding or grafting, grow on in a nursery bed for two or three years.

SOME POPULAR SPECIES	
S. aria *(Europe)* (Whitebeam) Round-headed tree with oval to pear-shaped leaves, downy and silvery-white when young, later becoming green. Golden autumn colour. Flattened heads of creamy-white flowers in late spring and early summer, followed by round, scarlet fruits in early autumn. Medium. **S. aucuparia** *(Europe, Western Asia)* (Mountain ash, rowan) Erect when young, becoming spreading and more graceful with age.	Pinnate leaves turning orange and yellow in autumn. White hawthorn-like flowers in late spring and early summer. Large bunches of round or slightly oval bright red berries. There are several good varieties. Medium. **S. vilmorinii** *(Western China)* Large shrub or small tree with wide-spreading habit. Dark green pinnate leaves. White flowers in early summer followed in autumn by red fruits that gradually become almost white. Small tree.

Sorbus aucuparia

SPARTIUM

Spartium
Spanish broom
SH/O/FS

A genus of a single species. A hardy, deciduous shrub with green stems that help to compensate for its otherwise rather sparse appearance out of flower.

How to grow
It needs a light, well-drained soil and full sun. Does not do well on limy soils.

Remove faded flowers to prevent seeding. In early spring cut back the previous season's growth to within about 5 cm (2 in) of the old wood.

Propagation
Seed is the main method of propagation. Sow in early or late spring after soaking in tepid water for 12 hours. Sow two seeds to each 8 cm (3 in) pot and thin to leave the strongest one. Germination will take about six to eight weeks. Pot on in early summer, using a loam-based compost, and plunge the pots in the frame. Remove the lights (tops) about a month later, and plant out in autumn.

SOME POPULAR SPECIES	
S. junceum *(Southern Europe, Canary Islands)* Gaunt shrub with rush-like green stems and a few narrow, mid green leaves.	Fragrant, golden-yellow pea-type flowers in summer on slender, almost leafless stalks. Medium.

Spartium junceum

Spiraea thunbergii

Spiraea
SH/O/FS

Hardy, deciduous flowering shrubs, taking their name from *speira*, an ancient Greek name for a wreath or a plant used for garlands.

How to grow
Spiraeas succeed in fertile soil in full sun.

Spring-flowering species such as *S.* x *arguta* and *S. thunbergii* should be pruned after flowering, removing wood that has flowered but retaining as many young shoots as possible to flower the following year.

Summer-flowering species such as *S.* x *bumalda* and *S. japonica* are best cut back to within 7.5–10 cm (3–4 in) of the ground in late winter or early spring.

Propagation

Semi-ripe cuttings root without much difficulty. Insert in containers under a cold frame in mid or late summer. Pot up singly when rooted (usually it takes about a month). Overwinter under the frame, pot on in mid spring, then harden off and plunge outdoors to grow on for a couple of years before finally planting in autumn.

Hardwood cuttings root in about a year if inserted in sandy soil under a cold frame or in a sheltered spot outdoors. These are best taken in late autumn. The method is described later in the book on page 106.

Suckers can be separated from the parent shrub between mid autumn and late winter, and replanted immediately.

SOME POPULAR SPECIES			
S. x arguta *(Garden origin)* (Foam of May, bridal wreath) Rounded, bushy, shrub with masses of white flowers on arching stems in mid and late spring. Medium. **S. x bumalda** *(Garden origin)* Almost like an herbaceous plant in habit and	appearance. Flat heads of deep pink flowers on erect stems in mid and late summer. There are several varieties, including 'Anthony Waterer' (bright crimson flowers; leaves often flecked pink and cream), 'Goldflame' (young foliage variegated gold). Small.	**S. japonica** *(Japan)* Erect, open habit. Flattened heads of small pink flowers on erect stems in mid and late summer. There are several varieties. Small. **S. thunbergii** *(China, Japan)* Bushy, twiggy shrub with arching stems. Small	clusters of white flowers on bare stems in early and mid spring. Medium. **S. x vanhouttei** *(Garden origin)* Graceful, arching stems densely covered with lobed, grey-green leaves. Pure white flowers in late spring and early summer. Small to medium.

Symphoricarpos

Snowberry

SH/O/SD–FS

Berry-bearing shrubs, the name being derived from *symphoreo* (to bear together or accumulate) and *karpos* (fruit), alluding to the clustered berries. You will sometimes see the generic name spelt *Symphoricarpus*. Easy plants that grow well in any reasonable soil, in sun or shade.

How to grow

Thin out overgrown plants or unwanted shoots in early spring. Unwanted suckers may need to be removed occasionally—best done in winter.

Propagation

Seed is little used, mainly because cuttings and suckers are so easy.

Hardwood cuttings taken in mid or late autumn and inserted in sandy soil in a sheltered position outdoors will root and be ready for planting out a year later.

Well-rooted suckers can be separated from a healthy clump in mid or late autumn and replanted immediately in their final positions.

SOME POPULAR SPECIES	
S. albus *(Eastern North America)* (Snowberry) Erect, clump-forming shoots with pale to mid green broadly ovate leaves. Plump white berries from mid autumn lasting into mid winter. *S. a. laevigatus* (which may be sold as *S. rivularis*) forms dense thickets and has heavy crops of large berries. Medium.	**S. x chenaultii** *(Garden origin)* Dense-growing shrub with small, round, pinkish-white berries, purplish-red on the exposed side. 'Hancock' is dwarfer. Small. **S. x doorenbosii** *(Garden origin)* A useful, non-suckering hybrid, raised in Holland. Several varieties. Medium.

Symphoricarpos albus laevigatus

Syringa
Lilac
SH/AK/FS

A genus of 30 species of hardy deciduous shrubs and small trees grown for their flowers, which are usually very fragrant. The common lilac can become very leggy and sparse at the base with age, but is nevertheless a magnificent shrub at its best.

How to grow
Lilacs do best on chalky ground, but they will grow in any reasonable soil. Feed or mulch each spring.

Remove faded flower heads, and cut out any suckers that appear, taking them back as close as possible to the roots or main stem.

The following year's flowers will be carried on vigorous shoots developing even while the shrub is in flower, so be careful not to remove these—in fact no regular pruning is necessary. An old bush may need rejuvenating, however, in which case it can be cut down in winter to about 60–90 cm (2–3 ft) above the ground (the bush will take two or three years to flower again).

Propagation
Seed is slow, and not suitable for raising the named varieties. Cuttings are easier and more reliable.

Cuttings of semi-ripe wood provide the quickest and easiest means of increase. Take them in mid summer, with a heel, and root in warmth at about 16°C (60°F). Pot up when rooted (usually it takes about six weeks), over-winter under the frame, and pot on in mid spring. Harden off and plunge outdoors to grow on for two or three years, lifting and repotting each spring.

Budding in mid summer, using privet as the rootstock, is considered to produce long-lived plants, but it is difficult. Grow on for a couple of years before planting out.

Syringa 'Firmament'

Syringa xjosiflexa 'Bellicent'

SOME POPULAR SPECIES

S. x josiflexa *(Garden origin)*
Deep green ovate, taper-pointed leaves. Large terminal clusters of fragrant pink flowers in late spring and early summer. 'Bellicent' has particularly large panicles of clear pink flowers. Medium.

S. x prestoniae *(Garden origin)*
A group of hybrids with large, erect or drooping panicles in early summer. Varieties include 'Audrey' (deep pink), 'Elinor' (purplish-red in bud, opening pale lavender), and 'Isabella' (purple). Medium.

S. vulgaris *(Eastern Europe)*
Upright shrubs with heart-shaped leaves and pyramidal heads of fragrant lilac-coloured flowers in late spring and early summer. Mostly grown as the many varieties, such as 'Charles Joly' (double, dark

purplish-red), 'Congo' (single, red in bud opening pink), 'Firmament' (lilac blue), 'Primrose' (single, pale yellow), 'Souvenir d'Alice Harding' (double, creamy-white flowers), and 'Souvenir de Louis Späth' (single, wine red). Medium to large.

Ulex

Gorse, furze, whin

SH(E)/O/FS

Spiny evergreen flowering shrubs with yellow pea-type flowers. Only one species is in general cultivation, a plant that thrives on poor, dry, heath or downland in the wild.

How to grow
Best on poor, dry, preferably acid soil in full sun, but buy young container-grown plants and water if dry until they become established.

Gorse tends to become leggy at the base, in which case it can be cut back to within 15 cm (6 in) of the ground in early spring—this will stimulate the production of new growth from the base of the plant.

Propagation
Seed sown in early or mid spring is the main method of propagation. Soak the seed in tepid water for 12 hours before sowing two seeds in each 8 cm (3 in) pot. Germinate under a cold frame or in slight warmth (ideally 13–16°C/55–60°F).

As soon as the seedlings show, remove the weaker one in each pot. The young plants should be ready to harden off in early or mid summer, when they can be plunged outdoors to grow on until autumn when they are finally planted out.

Cuttings can be tricky, but must be used to increase double-flowered forms. In mid or late summer, prepare 8 cm (3 in) semi-ripe cuttings

Ulex europaeus

and insert singly in pots under a cold frame, where they should be overwintered. If rooted, pot on in mid spring. Harden off in early summer and plunge outdoors for the summer, ready to plant in autumn.

SOME POPULAR SPECIES

U. europaeus *(Western Europe)*
Spiny, densely-branched shrub with pea-type yellow, fragrant flowers in spring, and intermittently

into autumn and winter. 'Plenus' is one of the best forms for the garden. It has masses of semi-double flowers in mid and late spring. Medium.

Viburnum (deciduous)

SH/O/FS

A genus of deciduous and evergreen shrubs. For evergreen species see the next entry.

Some of the deciduous species flower in winter on naked wood, which makes them especially useful for winter interest. But these put on a very poor display in comparison with the summer-flowering species.

How to grow
Viburnums are not fussy about soil, and most do well on chalk. Pruning is not normally necessary.

Propagation
Seed is slow—germination may take up to two years. Taking cuttings is an easier method and more widely used.

Prepare semi-ripe cuttings with a heel in mid or late summer. Root in warmth (16–18°C/60–65°F if possible), or in a cold frame. Pot up when rooted (it usually takes a month or two), and overwinter in a cold frame. Pot on in spring, harden off, and plunge outdoors in late spring. Grow on for two or three years before final planting.

Layering is easy and reliable—see evergreen species.

Viburnum opulus

SOME POPULAR SPECIES	
V. x bodnantense *(Garden origin)* Upright habit. Clusters of fragrant white flowers, flushed pink, on naked branches in winter. 'Dawn' and 'Deben' are improved varieties. Medium to large. **V. carlesii** *(Korea)* Rounded shrub with broadly ovate dull green leaves and clusters of fragrant white flowers in spring. Small. **V. farreri** *(V. fragrans)* *(North China)*	Wide-spreading shrub with bright green leaves, tinted bronze when young. Pendent clusters of fragrant white flowers, pink in bud, in winter. Medium to large. **V. opulus** *(Europe)* (Guelder rose) Erect, thicket-forming shrub with maple-like dark green leaves. Heavily scented white flowers in late spring and early summer, followed by persistent red fruits. Many good varieties.

Viburnum (evergreen)

SH(E)/O/PS–FS

Deciduous species have been dealt with in the previous entry. The evergreens lack the bold flowers of some of the deciduous species, but they include some invaluable shrubs, including *V. tinus*, which will flower all winter, and will do well even in shade. Both evergreen and deciduous viburnums are useful as either informal hedges or screening plants.

How to grow
As deciduous species.

Viburnum tinus

Propagation
Seed is slow, but cuttings are not difficult—see deciduous species.

Layering is reliable and ideal where just a few plants are required. Peg down shoots two or three years old in early or mid autumn. These will usually be ready to lift in one or two years. Replant immediately in a nursery bed, or pot up. Grow on for a year before planting in final positions.

SOME POPULAR SPECIES			
V. x burkwoodii *(Garden origin)* Ovate, dark green leaves. Clusters of fragrant white flowers, pink in bud, in spring. Medium. **V. davidii** *(China)* Low, compact shrub with leathery dark green leaves, prominently veined. Flat	heads of white flowers in early summer, followed by very attractive bright turquoise berries if male and female plants are present. Small. **V. rhytidophyllum** *(Central and Western China)* Spreading shrub with	stout branches and lance-shaped, deeply wrinkled dark green leaves. Flat heads of white flowers are carried in late spring and early summer. Red fruits becoming black in autumn (but not likely if only one specimen is planted). Large.	**V. tinus** *(South East Europe, Mediterranean region)* (Laurustinus) Dense shrub with ovate to lance-shaped leaves. Pink-budded white flowers in flat heads from late autumn to late spring. 'Eve Price' is more compact.

Vitis

Ornamental vine

CL/AK/PS–FS

A genus of about 70 species of evergreen or deciduous climbers. The species most widely grown, and included here, are deciduous. They are best grown on a pergola or on a trellis; they cling with tendrils and need some form of support.

How to grow
Will grow in any well-drained soil, and do well on chalky ground.

Pruning consists of cutting out unwanted growth in summer.

Propagation
Relatively easy from seed. Sow in containers under a cold frame in late autumn, then move into warmth (16°C/60°F) in late winter. Prick out singly when large enough, usually within about six weeks. Grow on in a cold frame for a year, potting on as necessary. Harden off in late spring, and stand outdoors until autumn, when they can be planted.

Hardwood cuttings root readily if taken in late autumn and inserted under a cold frame. Leave undisturbed for 18 months, then lift and pot up those that have rooted. Return to the frame, harden off in late spring, and grow on outdoors until the autumn, when they can be planted.

Layers will usually root within a year. Peg them down in autumn. Pot up the rooted plants and overwinter under cover. Harden off in late spring and treat as cuttings.

SOME POPULAR SPECIES	
V. coignetiae *(Japan)* Vigorous climber with thick, roundish but often slightly lobed leaves up to 30 cm (1 ft) across. These turn brilliant scarlet and crimson in autumn. The black berries have a purplish bloom. Very vigorous.	**V. 'Brant'** *(Asia Minor, Caucasus)* A hybrid vine involving the grape, and bearing edible black grapes. It is, however, grown for its lobed leaves that turn shades of dark red and purple in autumn. Moderately vigorous.

Vitis 'Brant'

Weigela

SH/O/PS–FS

A genus of 12 hardy deciduous shrubs, named in honour of Christian Ehrenfried von Weigel (1748–1831), a German professor of botany. Most of the plants grown are hybrids.

How to grow
For best results, grow them in a moist, rich soil, although they tolerate chalk well.

Weigela florida 'Variegata'

Remove a couple of old stems on mature plants as soon as flowering is over. Cut back shoots with faded flowers.

Propagation
Cuttings are used almost exclusively.

Hardwood cuttings are easiest. Take them in mid or late autumn and root in sandy soil under a cold frame, or in a sheltered position in the garden. Remove frame lights (tops) in summer. Lift the rooted cuttings in autumn and plant out in a nursery bed (or pot up) and grow on for another year before planting out.

Semi-ripe cuttings taken in mid summer will root readily in a propagator. Pot up when rooted (it usually takes five or six weeks) and overwinter in a cold frame. Pot on in spring, then harden off and plunge outdoors. Plant out in autumn.

SOME POPULAR SPECIES	
W. florida *(China)* Wide-spreading shrub with arching branches bearing oval to lance-shaped leaves. Pink funnel-shaped flowers in late spring and early summer. 'Foliis Purpureis' has purple-flushed leaves, 'Variegata' has leaves edged creamy-white. Medium.	**Garden hybrids** There are dozens of hybrids with bolder flowers and a greater colour range. Popular ones include 'Abel Carriere' (rose-carmine), 'Bristol Ruby' (ruby-red), and 'Newport Red' (red). There are also whites, such as 'Mont Blanc'. Small to medium.

Wisteria

CL/O/FS

Caspar Wistar (1761–1818), Professor of Anatomy in the University of Pennsylvania, is the man who gave his name to these most magnificent climbers. They look superb as wall-trained plants, or growing over an arch or pergola.

How to grow
Wisterias need a moist, fertile soil to do really well, but will put in a respectable performance on most reasonable soils.

Wisteria sinensis

Prune annually in mid or late summer, cutting back the current year's growth to about 15 cm (6 in).

Propagation

Seed is slow. Cuttings are easier and quicker, and layering easiest.

Semi-ripe cuttings taken in mid or late summer will root in a propagator, but can be fickle. They should be ready to pot up in about five or six weeks. Overwinter in a cold frame, then harden off and plunge outdoors to grow on for another couple of years before planting.

Layer young stems in spring, first kinking or slitting the stems. They should be ready to lift in the following spring. Grow on for a couple of years before finally planting out.

SOME POPULAR SPECIES	
W. floribunda *(Japan)* Twining climber with pinnate leaves and drooping bunches of pea-type fragrant, violet-blue flowers in late spring and early summer. These are often 25–30 cm (10–12 in) long. 'Alba' has white flowers tinted lilac, 45–60 cm (½ to 2 ft), 'Macrobotrys' has very long cascading bunches of blue flowers up to 1 m (3½ ft) long. Vigorous.	**W. sinensis** *(China)* Similar to last, but the stems twine anti-clockwise, unlike the previous species. 'Alba' is white, 'Plena' has double lilac flowers. Very vigorous. **W. venusta** *(Japan)* White trusses with large fragrant flowers, late spring and early summer. Downy young foliage and shoots. Vigorous. *Violacea* has violet-coloured flowers.

Yucca

SH(E)/O/FS

Evergreen trees and shrubs native to southern USA, Mexico, and Central America. They are distinctive plants in leaf and flower, and for that reason do not mix well with other shrubs. They are better as isolated specimens.

Yucca flaccida

How to grow

Will do well in any ordinary, well-drained soil, even if poor and sandy. Not suitable for heavy clay soils. No pruning is necessary.

Propagation

Seed is slow and fickle, and not often used. Suckers are much easier.

Remove rooted suckers in spring. Pot up individually and place in a cold frame to grow on for a year. Then in late spring plunge the pots outdoors to grow on for another year before planting out.

SOME POPULAR SPECIES	
Y. filamentosa *(South East USA)* Stiffly erect sword-like glaucous green leaves, arising from a rosette. Flower spikes are produced on relatively young plants. The creamy-white bell-shaped flowers are produced on 1.8 m (6 ft) erect, stiff stems, mid or late summer. Small to medium. **Y. flaccida** *(South East USA)* Tufts of long, lance-like leaves, the end of each one	bent down and the edges with curly white threads. Creamy-white flowers on erect spikes 60–120 cm (2–4 ft) tall in mid or late summer. Small. **Y. gloriosa** *(South East USA)* (Adam's needle) Distinguished from *Y. filamentosa* by having a stem bearing rosettes of stiff, sword-like, spine-tipped leaves. Creamy-white bell-shaped flowers on erect stems up to 1.8 m (6 ft) long in autumn. Small to medium.

Terms & Techniques

This section includes an explanation of technical and horticultural terms used in the first part of the book, as well as some that you might encounter in other tree and shrub books.

It is more than just a glossary, however. Where relevant, the practical implications are explained, and for practical tasks the jobs are illustrated with step by step illustrations where these are helpful to the text.

We have not included specific chemicals for pest and disease control because these vary from country to country, and new ones are continually appearing, sometimes to replace those that are currently in use.

If you familiarise yourself with this part of the book you should find it a useful aid to propagation and tree and shrub care as well as a source of explanation for those terms that you may not be quite sure about.

Acid compost/soil

A compost or soil with a low pH (*see* pH), necessary for those plants (known as acid loving) affected by too much lime in the soil. *See* Chlorosis, Compost, and Ericaceous compost.

Air layering

Many trees and shrubs can be propagated by conventional layering, but sometimes it is not practical to peg down suitable stems and then air layering can be considered. Air layering does not depend on the stem being brought into contact with the ground.

Among the trees and shrubs for which air layering can be a useful method of increase are the many kinds of hamamelis, lilac, magnolia, and rhododendron.

A couple of inches below a leafy part of the plant, make a slanting cut with a sharp knife, being careful not to slice through the stem. If you are worried by this, simply remove a strip of 'bark' about 1 cm (3/8 in) wide right round the stem. Dust or paint with a rooting hormone.

Pack damp sphagnum moss or peat around the stem. Tie a piece of

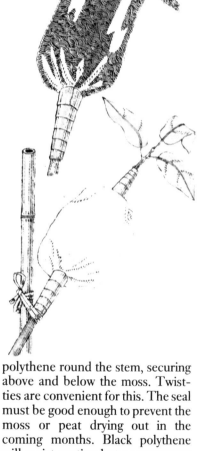

polythene round the stem, securing above and below the moss. Twist-ties are convenient for this. The seal must be good enough to prevent the moss or peat drying out in the coming months. Black polythene will assist rooting but a transparent sheet may enable you to see roots developing. One solution is to use a second layer of black polythene, which you can remove to check on progress.

Support the layer by tying the branch to a cane, so that it does not snap off in a wind.

When roots have formed (it could take as little as a month or as long as a year), sever the plant and pot it up.

Aphids

A group of insects of which greenfly and blackfly are the most common. These attack a wide range of trees and shrubs, although they are not likely to be as much of a problem on 'softer' plants. Fortunately they are easily controlled by a wide range of insecticides.

Ascending
Branches or stems neither prostrate nor upright, but growing obliquely.

Balled plant
A tree or shrub grown in a field and lifted with a ball of soil that is then wrapped in hessian or a plastic material to keep the soil around the roots. Conifers and evergreens such as hollies are sometimes sold this way. *See also* Bare-root plant and Container-grown plant.

Bare-root plant
A field grown tree or shrub lifted in the dormant season without a ball of soil. Bare-root plants are now much less common than they used to be, but will transplant successfully if planted at the right time.

Berry
Botanically a fruit in which seeds are protected only by a fleshy wall formed from the ovary. But the term is generally used, and in this book, loosely to cover other forms of fruit, such as those of hawthorns and cotoneasters.

Bicolor
Two-coloured. Usually applied to flowers in which one colour is contrasted with another.

Bigeneric
Describes a type of hybrid derived from a cross between two different genera. For instance *Laburnocytisus* is a bigeneric hybrid between *Laburnum anagyroides* and *Cytisus purpureus.*

Bipinnate

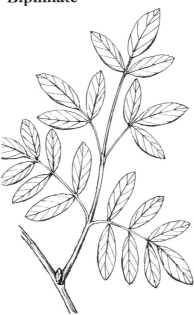

Term applied to leaves composed of several separate segments, which are themselves divided into separate segments.

Botrytis
Popularly known as grey mould, this fungus disease will not be a problem on established trees and shrubs, but it may affect cuttings. The fungus usually starts on dead or dying tissue (dead or dying leaves for instance), but if left unchecked can spread to affect healthy tissue. The common name of this disease is apt—the affected part becomes enveloped in a grey mound of fluffy mould. If moved or disturbed, dust-like spores will probably fly up and drift to start new infections.

Many modern fungicides will control botrytis if you treat the plant early. It is worth dipping your cuttings in a solution of fungicide as a routine precaution.

Bowl
The clear trunk of a tree between ground level and the first branch.

Budding
Budding is a method used by professionals to propagate some plants that are either difficult to root from cuttings or layers, or that do not do well on their own roots (the rootstock will largely control the vigour of the plant).

You can try it yourself if you have a suitable rootstock—usually plants growing in open ground from which you remove all branches and leaves from the bottom 30 cm (1 ft) of the plant in mid summer. For the bud, choose a stem of similar diameter if possible—discard any soft growth, remove any leaves flush with the stem, and make sure there are some well matured buds.

There are two main methods of budding—**T-budding** and **chip-budding**.

T-budding. Close to the base of the plant make a T-shaped cut and loosen the flaps of bark.

103

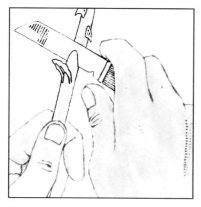

Using the variety that you want to bud, make a *shallow* cut around the bud, towards the tip of the stem— starting about an inch above the bud and finishing an inch below.

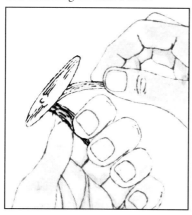

Remove the small piece of wood from the back of the bud by bending the bark backwards.

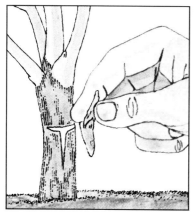

Insert the bud, tail up, between the flaps in the rootstock, then trim the tail flush.

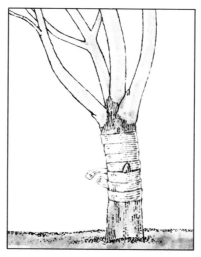

Bind the bud in place with raffia or tape, making sure the bud is left exposed, or use a budding patch. If the bud takes, cut off the rest of the rootstock in late winter, just above the bud.

Chip-budding is generally easier. Make a 6 mm (¼ in) deep cut into the rootstock, then starting about 2.5 cm (1 in) above it cut out a wedge or chip. Cut out a similar wedge from the budwood, making sure the bud is in the centre of the chip.

Make sure the chip is a good fit and in close contact, then bind it with clear grafting tape, making sure the edges of the chip are well sealed. *Remove the tape after three or four weeks.*

In winter cut the rootstock back to the bud.

Calcareous
Containing chalk or lime. A calcareous soil contains chalk or lime.

Calcifuge
A plant that dislikes lime or chalk, such as rhododendrons and most heathers.

Callus
Thickened tissue that forms over a wound.

Calyx (plural calyces)
The outer, protective part of a flower, consisting of a ring of usually green modified leaves (the sepals) that are fused together at the base to form a bowl, funnel, or tube.

Cambium
Usually taken to mean the very narrow layer of active tissue between the bark and wood. The cambium of a rootstock and scion or bud must be in contact with each other for a graft or bud to unite.

Catkin

A kind of flower spike, usually unisexual and often pendulous though sometimes erect. Composed of scale-like bracts, surrounding stalkless flowers usually without petals. Birches, hazels, and willows are among the well-known plants with catkins.

Clone
A group of identical plants produced vegetatively from one original parent.

Chlorosis

The loss of, or lack of, chlorophyll (the green colouring matter in leaves) usually due to the lack of certain available elements in the compost necessary for its production. The leaves become bleached or yellow.

Plants that prefer an acid compost may become chlorotic if the compost contains too much lime, which prevents these plants from obtaining some essential nutrients.

Chlorotic plants can be treated with a chelated compound (Sequestrene).

Cold frame

Sometimes called garden frame. A structure offering weather protection—traditionally made from wood and glass but increasingly now from aluminium, metal, and plastics—though glass is still popular for the glazing.

Frames with glass sides have the advantage of better light penetration, but are colder in winter unless insulated then.

Frames are usually unheated, but soil or air warming cables can be used.

Unless otherwise stated, cuttings can be inserted directly into good soil or compost spread over the base.

Frames have been suggested for propagating most of the trees and shrubs in this book. But you will not succeed if the plants are allowed to dry out—and those in containers are particularly vulnerable. *Plants in containers in a frame will need almost as much regular attention as those in a greenhouse.*

Compost

The compost is crucially important for pot-grown plants. Earlier generations of gardeners used to have favourite recipes with ingredients that most of us would find practically impossible to obtain now: leaf-mould, quality fibrous loam, crushed bones, decomposed sheep manure, mortar rubble and crushed brick, and even crushed potsherds, are examples. Fortunately, most plants will grow well in one of the modern standard potting composts, whether loam based or peat based. The important thing is to grow on young plants in a proper potting compost and not ordinary garden soil, if you want to give them a good start.

Conifer

A tree that bears cones. Pines, firs, and cedars are examples.

Container-grown plant

A plant grown and sold in a container, usually a rigid or flexible plastic pot. Container-grown plants suffer much less root disturbance than bare-root or balled plants (*See* Balled plant, Bare-root plant), transplant better, and enable you to plant at any time of the year if the ground is not frozen or waterlogged.

Beware of *containerised* plants—plants potted up shortly before sale, as these will be little better than bare-root plants. Lift the plant by the stem—if established in the container it will not pull out of the compost.

Corolla

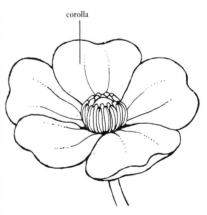

corolla

Term applied to that part of the flower made up of true petals, usually forming a conspicuous inner whorl, backed or surrounded by the usually green sepals (calyx).

Corymb

Term used to describe a flat-topped cluster of flowers, composed of a series of flowers borne on individual stalks, each of different length. These form a rounded and more or less level head of flowers.

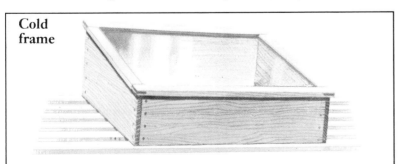

Cold frame

Crotch
The point where the main trunk divides into branches.

Crown
Usually refers to the part of the plant just below or at soil level, from which the shoots grow.

Cultivar
This is the botanist's name for a variety raised in cultivation. Some varieties of a plant occur naturally, and these are regarded as botanical varieties and should be printed in italics, with a lower case initial letter, after the species name. Most varieties occur in cultivation (sometimes by chance, often the result of a breeding programme), and these are strictly 'cultivars', which should be printed in Roman type, in single quotation marks, and with a capital initial letter.

To the gardener it makes no difference whether the variety occurred in a garden or in the wild (it makes the plant no more or less desirable), and to most gardeners 'variety' is more familiar and less pedantic so we have used that term throughout this book.

Cutting
The majority of trees and shrubs are propagated from either semi-ripe or mature (hardwood) cuttings.

Semi-ripe cuttings use shoots produced in the current season that are starting to become harder as growth slows down. Semi-ripe cuttings usually mean the current-season's growth taken in late summer.

Unless otherwise stated, make the cutting 10–15 cm (4–6 in) long, first removing the tip if this is soft.

Trim off the lowest leaves, and insert the cutting. If necessary dip the end in a rooting hormone first.

Pot individually or space the cuttings about 8 cm (3 in) apart. Water in—using a fungicidal solution if the cuttings have not been dipped in one.

Hardwood cuttings are taken during the dormant season, using fully mature stems.

Unless otherwise stated, make the cutting 15–25 cm (6–10 in) long.

Make a sloping cut just above the top bud and cut across horizontally at the base, which is best dipped in a hormone rooting preparation.

Insert the cuttings with about three buds just above soil-level (most of the cutting should be in the soil). If growing a single-stemmed plant, such as a tree, plant deeper so that the top bud is just below the surface (the increased depth will inhibit the development of the lower buds).

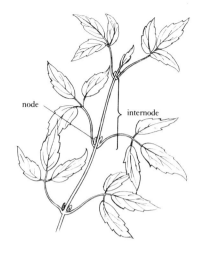

node

internode

Internodal cuttings are not often used, but clematis are usually propagated this way. Each cutting should have a node (with the pair of buds) and 2.5–5 cm (1–2 in) of stem below it. Thin wood from one-year-old plants will root more readily than from older plants.

Rooting in containers. Use an all-purpose compost, or a seed compost (vermiculite and perlite are also very successful) for rooting cuttings; potting compost may contain a harmful level of nutrients.

A propagator will help to root the more difficult kinds, and mist propagation will help with the really tricky ones. You can, however, provide the necessary humidity with the aid of a polythene bag over the pot, but avoid direct sunlight, which could overheat or scorch the plants.

Shade cuttings from direct sunlight until they have rooted and are growing strongly.

Deciduous

A plant that loses its leaves at the end of the growing season (in other words, it is not evergreen). It is normally used in the context of trees and shrubby plants, rather than herbaceous plants.

Deciduous trees and shrubs should be transplanted when dormant (unless container-grown), roughly from mid autumn to early spring.

Decumbent

Stems that lie on the ground for part of their length, then turn upwards.

Dentate

Coarsely-toothed, usually used to describe the edge of a leaf.

Die-back

A term used to cover several different diseases, all of which cause the shoot to die back from the tip. Affected branches should always be pruned back to completely healthy tissue. A wound paint is usually used to cover the wound after pruning, though there is some debate about the use and effectiveness of these.

Dioecious

Term used to indicate that all the flowers on any one plant are either entirely female or entirely male.

Division

Division is a self-explanatory term—it is simply a matter of dividing the old plant.

Not all plants are suitable. Only those with a fibrous root system and a crown that can be divided into sections containing shoots and roots can be propagated this way.

Dormant period

A temporary period when the plant ceases to grow, usually but not always coinciding with winter.

Ericaceous compost

An acid compost containing very little or no lime. Ericaceous plants (those belonging to the Ericaceae or heather family) are generally lime-haters, but plants belonging to other families may also dislike lime.

You can buy both loam-based and peat-based ericaceous compost mixes, though they are less widely available than ordinary composts.

A lime-free compost should be used for raising seedlings or propagating cuttings of lime-hating plants.

Evergreen

A plant that does not shed its leaves all at the same time, like a deciduous tree or shrub. Although the leaves are shed eventually as new ones grow, this is done gradually so the plant always appears clothed.

Evergreens are useful for their year-round interest, but too many can give the garden a 'heavy' appearance, and plenty of deciduous trees and shrubs are also needed to add variety and changing interest to the garden.

Family

One of the groupings into which botanists place plants. All plants within a family will have some common characteristics that are not found in other families. The family name ends in 'ae': Rosaceae, Ericaceae, for example.

Fastigiate

Erect, upright growth habit. Fastigiate trees can be particularly useful for a confined space. Well-known fastigiate trees are the Lombardy poplar and *Prunus* 'Amanogawa', though there are fastigiate varieties of many other popular trees.

Feathered

A term used in connection with a maiden (usually year-old) or young standard tree with lateral shoots on the main stem. The 'feathers' (side branches) are usually removed after a year or two to build up a strong, thick bole.

Feeding

All trees and shrubs will benefit from an annual dressing of a balanced general fertiliser for the first few years after planting. Apply this in spring and rake it in lightly. Established ornamental shrubs and trees should not need regular feeding unless you suspect a particular deficiency.

Adding bonemeal at planting time is a good idea.

Flore-pleno

A term that indicates doubleness, though the plant may be semi-double.

Floret

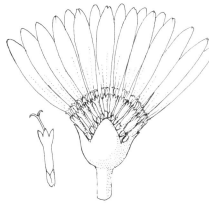

A single flower that forms part of a large head. In the daisy family (Compositae), each flower is really a head of tiny but closely-packed florets arranged around a central disc.

Foliar feeding

Some fertilisers are sold specifically as foliar feeds, but many ordinary liquid (and soluble powder) fertilisers can also be used as a foliar feed (the plants absorb the nutrients through the leaves). Generally it will do just as much good down at the roots, but sometimes an ailing plant or one that seems slow to grow away after planting may benefit from a foliar feed.

Frost protection

Most of the trees and shrubs in this book are very hardy, but some of them are suitable for only very mild areas, while others will tolerate a moderately severe winter but not a very bad winter. You can help these plants by giving them a sheltered position, ideally near a wall that will protect them from the prevailing cold wind, and by making sure they have additional protection while young.

The most vulnerable are re-cently-planted trees and shrubs. Evergreens can be particularly vulnerable, mainly because they will lose moisture through their leaves, especially in high winds, but be unable to replace it because the roots are not established. A simple polythene screen fixed to canes may be all that is necessary.

The stems of very vulnerable plants can have hessian and straw tied around the trunk, but this is unlikely to be necessary for any of the trees in this book. Generally, if you live in a very cold area it is better to confine your choice to plants that are very hardy—there are plenty of them.

Fungicide

A chemical that will kill or control fungus diseases (or at least some of them). Some have a systemic action (although it may not be as effective as spraying the leaves).

Gall

An abnormal outgrowth sometimes seen on trees and shrubs. A well-known one affects oaks, forming the so-called oak apples.

Galls are the result of irritation set up by insects or bacteria. Sometimes galls can be as big as a football, but they are usually more the size of a marble or smaller. They seem to have little or no effect on the plant.

Genus (plural genera)

A group of allied species. The genus is the first word of a plant name, and has a capital letter when written as part of a full name: in *Berberis darwinnii*, *Berberis* is the genus, *darwinii* the species (*see* Species).

The genus is equivalent to a Surname, the species to the individual and unique member of the family. A genus may contain only one species, or it may contain more than a thousand. It depends on how closely botanists consider various plants are related.

Germination

The emergence of a new plant from a seed. *See* Seeds, plants from.

Glabrous

Smooth, though strictly it means hairless and usually refers to a part of a plant that is hairless.

Glaucous

Bluish-grey; covered with a 'bloom'. Usually used to describe leaves or stems.

Grafting

Grafting is the joining of two plants so that they unite and grow as one plant. Grafting is sometimes used to control the plant's rate of growth (the rootstock often determining vigour), but grafting also provides a means of propagating those plants that are difficult to raise from cuttings or that will not come true to type from seed.

Grafting is quite a complex job, and you need to start by planting young plants for the rootstocks and letting these grow on for a season.

In mid winter cut off vigorous hardwood stems from the plant that you want to propagate, to provide the scion material. Bundle these together and insert the bottom 15 cm (6 in) into the ground in a cool, well-drained position.

Whip-and-tongue graft Prepare the rootstock just before the leaf buds break, by trimming all shoots off the bottom 30–38 cm (12–15 in) of the stem.

Then cut back the rootstock to the point at which you want to graft, making a sloping cut at the top about 4 cm (1½ in) long.

Lift the scion wood and cut across a stem of matching thickness, just above the fourth or fifth bud from the base. Then make a slanting cut at the end, at the same angle and size as the one on the rootstock, ending the cut just below the bottom bud.

Carefully make a slit in both rootstock and scion (see above and below) to produce lips that can be interlocked when pushed together.

The lips will hold the scion to the rootstock. Bind the graft with

plastic grafting tape, and seal the top of the scion with a bituminous tree paint.

Remove the tape once the cut surfaces start to callus.

Saddle grafting Prepare the rootstock and scion as shown.

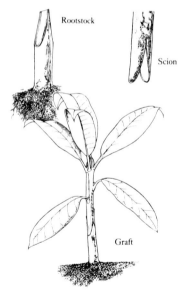

Rootstock

Scion

Graft

Side wedge grafting This is the easiest method if you are not used to grafting. Make two sloping 4 cm (1½ in) cuts on the scion, and a cut of the same length on the rootstock, about 5 cm (2 in) above the ground. Bend the rootstock slightly while you insert the scion, then bind the graft with grafting tape. Remove the tape and cut back the rootstock once the graft has taken.

Grey mould
A disease characterised by a grey, fluffy growth of mould (*See* Botrytis).

Gumming
Plums and cherries, including the ornamental forms, sometimes exude a resin that is gum-like in appearance, especially if the bark is injured.

It normally does no harm, but if a lot of rather cloudy gum is produced it may be a sign of infection by bacterial canker.

Half-hardy
Likely to be damaged by frost. Usually applied to bedding plants that can spend the summer outdoors but not the winter and must, therefore, be brought inside before the first frosts in autumn.

Hardening off
The process of acclimatising a plant to cooler conditions. Plants frequently receive a severe check to growth if moved suddenly from a warm atmosphere into a cold exposed one.

Cold frames are often used for hardening off greenhouse-raised plants, but plants raised in frames will themselves need hardening off. The frames are normally covered in cold weather, but as warmer conditions arrive the lights (tops) should gradually be left off during the day and then replaced each night. After a few weeks of this treatment it should be possible to leave the lights off at night too.

Hardy
Generally taken to mean a plant that will not be killed by frost. But there are degrees of hardiness, and the more severe and prolonged the frost the greater the number of plants likely to be damaged or killed.

Haw
Fruit of *Crataegus* (hawthorn) species.

Heading back

The process of cutting back a tree or shrub very severely, shortening some or all of the main branches. The term is often used in connection with grafting and means cutting back the rootstock to receive the graft, or after budding or side grafting.

Heel

A small strip of bark and wood that remains attached to a shoot used for a cutting when it is pulled away from the main stem. Normally the heel is trimmed back to shorten it before inserting the cutting.

Cuttings that do not require a heel are normally cut off with a knife.

Herbaceous border

In the strict sense, a border used only for herbaceous perennials. But many shrubs blend very well with normal herbaceous plants— *Perovskia atriplicifolia* and *Spiraea* x *bumalda* for instance.

Mixed border, containing both shrubs and herbaceous plants, are becoming more popular.

Honeydew

A sticky secretion left on leaves by insects such as aphids and whitefly. It can be a particular problem because an unsightly black mould often grows on it, and this looks disfiguring.

The solution is simple—control the insects responsible.

Hormone rooting preparations

Hormone rooting powders or liquids can be useful for rooting the more difficult plants, but many of the softer stemmed plants will root readily without assistance.

The two chemicals most widely used are naphthyl acetic acid (NAA) and indole butyric acid (IBA). There is some evidence that IBA is more effective on a wider range of plants, but both are useful.

Humus

The dark brown residue left when organic matter decays. The term is often used to describe the partly decayed brown, crumbly material such as well-made compost or thoroughly rotted leaves.

Hybrid

Usually a plant derived from crossing the two distinct species or (much less commonly) genera.

Sometimes a cross *within* a species is also described as a hybrid. But unlike an ordinary cross resulting from normal cross-pollination, which might produce an ordinary variety (technically, cultivar), these hybrids are likely to be the result of crossing two pure 'lines' (in other words stable varieties that breed true). The resulting cross is called an F1 hybrid, and is generally better than the contributing parents. F1 hybrids are not really applicable to trees and shrubs.

Inflorescence

The part of a plant bearing flowers. It is a term often used not so much in its botanical sense but to describe a flower head or spike that is unlike a typical flower (perhaps a spike where colourful bracts are the main feature, with the true flowers of secondary importance).

Indumentum

A dense covering of short hairs. The term is usually used to describe leaves or stems densely covered with short hairs, especially those of many rhododendrons.

Insecticide

A chemical for killing insects. There are many from which to choose, and most of them will be very effective against the insects that they are supposed to control. Sometimes, however, it is necessary to spray or dust more than once to achieve control. Always follow the manufacturer's instructions.

Systemic insecticides are translocated within the plant and can be useful against insects that are difficult to reach or control with contact insecticides.

Internode

The portion of stem between two nodes (joints from which leaves arise).

Juvenile foliage

Some trees and shrubs produce leaves of a different shape when they are young. Those of an ivy are less lobed when the plant has reached a mature, flowering stage. Leaves on a young eucalyptus are usually rounded and may clasp the stem, those on older trees are generally sickle-shaped and stalked.

Sometimes trees such as eucalyptus are coppiced or 'stooled' to stimulate a continuous supply of juvenile foliage.

Laciniate
Cut into narrow segments. Usually applies to leaves that are divided into fine segments.

Lanceolate
Term used to describe a lance-shaped leaf; one with a long, gradual taper.

Lateral shoot
Any shoot growing sideways from the main stem below the tip.

Layering
A method of propagation. There are several types of ground layering, apart from air layering (see page 102), but only simple layering is described here.

If you only want one or two plants and there are suitable low-lying branches that you can peg down from the parent plant, no prep-aratory work is necessary. If you want the maximum number of layers, low branches will have to be pruned on the parent plant at least a year ahead, to provide plenty of suitable shoots.

Cultivate the soil well where the plant is to be layered, incorporating peat and grit or coarse sand. Then trim the leaves and sideshoots off the chosen stem where it will be in contact with the ground. Mark the position where it will be pegged down (about 23 cm/9 in) behind the tip.

Take out a trench about 10–15 cm (4–6 in) deep with a straight back and a slope towards the parent plant. Peg the shoot down into this, first twisting or injuring the stem if recommended to encourage root-ing. Cover the layered stem with fine soil, leaving the leafy tip exposed. Water well and keep moist until well rooted.

The new plant can be severed from the parent once it is well rooted and ready for moving.

Layers often root more readily if the stem is injured, either by twisting, removing a small strip of tissue around the stem, or by making a small slit. If you do wound the plant, dust it with a hormone rooting powder.

Leader
The terminal shoot of a branch, which if left unpruned would continue to extend the line of growth either upwards or outwards.

A 'replacement leader' is a side growth that can take over the position of leading shoot once the leader has been cut out.

Leaf bud cutting
Leaf bud cuttings are not much used by amateurs, but the tech-nique can be useful for raising a lot of camellias for instance.

To make a leaf bud cutting, cut just above and below the node, leaving a leaf attached.

Pot up each leaf separately, just inserting the piece of stem into the compost. Given warm, humid conditions the cuttings should root and begin to grow.

Leaf curl
A disease that normally affects almonds, apricots, nectarines, and peaches (it is usually known as peach leaf curl), but also related ornamentals.

The fungus causes the affected leaf to become thickened and dis-torted, with dark red patches.

The usual control is to spray with a suitable fungicide just before leaf fall in autumn, then in mid and late winter, with a repeat application a fortnight later. Pick off and burn any affected leaves if at all possible.

Leaflet

A section of a compound leaf, usually resembling a leaf in itself.

Leaf-mould

A confusing term. It has two meanings. It can be rotted leaves, an ingredient of some traditional potting compost mixtures; or a general term applied to various fungus diseases that rot leaves.

Most leaf-rotting diseases can be controlled to some extent by modern fungicides.

Leaf spot

The name given to several fungal and bacterial diseases that cause discoloured patches on eaves.

Leggy

A term used to describe a plant that has become drawn and spindly, usually because of lack of light. The stem becomes elongated between the leaf joints, and the plant is generally weakened.

Linear

Long and narrow. Term used to describe leaves of this shape.

Loam

In many ways, the 'ideal' soil: neither wet and sticky nor dry and sandy. A good blend of clay, silt, sand, and humus. Good loam has a fibrous texture.

Loam is a basic ingredient of traditional potting composts, and is an ideal soil for the majority of trees and shrubs.

Loamless composts

Not all composts contain loam. Many are based on peat, but materials such as vermiculite may also be used. *See* Composts.

Long-day plants

These flower when the days are long and the nights short. They usually flower after they have been subjected to more than 12 hours of light each day for a period.

Lop

To remove or drastically shorten large branches. Done when pollarding trees, but not to be recommended, as lopping will spoil the symmetry of a tree.

Maiden

A tree or bush in its first year after budding or grafting.

Mealy bug

Easily identified pest. The young bugs are protected with what looks like a piece of cotton-wool. If only a few are present, pick them off or touch each of them with a paint brush dipped in alcohol. Otherwise use a systemic insecticide or spray with malathion.

Micropropagation

A scientific method of propagation, usually using a very tiny portion of the growing tip (though there are several techniques) and growing these on using nutrient gels in sterile conditions to provide a supply of plants. Sometimes micropropagation is used to provide a large number of plants commercially, sometimes as a method of providing healthy virus-free stock.

It is possible to buy the equipment and nutrients to do this at home, but unnecessary unless you want to try it for fun.

Midrib

The main rib that divides a leaf centrally along the length. It usually stands out more prominently on the back of the leaf than other veins and ribs.

Mildew

A white, powdery deposit on the leaves, sometimes spreading to the stem too. Some plants are more susceptible than others.

Pick off affected parts as soon as noticed. If the attack is severe it may respond to some of the modern fungicides, but the sooner an attack is treated the better the chances of success.

Mist

Misting a plant is a way of increasing humidity, which can be beneficial when rooting cuttings. It can be very successful *if done often enough*—even once a day may not be enough as the effect is fairly transient.

Use a fine mist, and do not spray plants with hairy leaves or those with fleshy succulent leaves.

Mist propagation

Mist propagation is widely used commercially—it can drastically improve the 'take' of some of the more difficult trees and shrubs being raised from cuttings. But if you need only a few plants it probably is not worth the expense of buying the equipment. And a high success rate is usually less important to an amateur, who usually only wants a few plants.

A mist propagator prevents the cuttings wilting by keeping them covered with a fine film of water. It does this by intermittently spraying them with a fine mist. Remember that you need to install a water and power supply (the system is likely to have a soil-warming system too).

Most cuttings will root more quickly, with a higher success rate, but a mist propagator is most useful for the plants that are normally quite difficult. These include: many acers, arbutus, deciduous azaleas, betula, camellia, corylopsis, cotinus, evergreen daphnes, elaeagnus, fothergilla, garrya, magnolia, pieris, many rhododendrons and wisteria.

Monoecious

A plant with separate male and female flowers *on the same plant* (but not in combined male/female flowers). The hazel is an example of a monoecious plant—the showy catkins are the males; the female catkins are small, red, and grow close to the stem.

Mulch

A thick dressing of garden compost, peat, or similar material applied around the plants on the surface of the soil. Pulverised bark is another popular mulching material, but even black polythene or gravel can be used.

Mulching is particularly worthwhile during the first few years while a tree or shrub is becoming established, as it will conserve moisture and reduce weed competition.

Any organic material such as compost, peat, or pulverised bark will need to be applied in a layer at least 5 cm (2 in) thick to suppress weeds.

Neutral soil

A soil with a pH of about 6.5, neither very acid nor very alkaline. *See* pH.

Node

A point on the stem from which leaves arise. In the axils of the leaves are buds from which new shoots can develop.

The space between nodes (leaf joints) is known as the internode.

Nursery bed

An area of ground set aside in which to grow on young plants until large enough to be planted in their final positions.

Obovate

Egg-shaped in outline, broadest at the tip.

Ovate

Egg-shaped in outline, broadest at the base.

Palmate

A compound leaf shape, with lobes radiating like the fingers of a hand.

Panicle

A branched flower cluster, each branch with numerous flowers on individual stalks, the youngest flowers at the top.

Pappus

Tuft of hair or bristles found in some flowers that later helps the seeds to become airborne.

Peat compost

Peat-based composts have become very popular in recent years. They can be produced to a consistent standard, and are light and not unpleasant to handle.

Lime is added to bring the pH up to an acceptable level, and nutrients to support plant growth. Sand or other ingredients may be used to produce a suitable structure and texture.

Do not assume that a peat-based compost will be suitable for lime-hating plants (unless of course it is described as an ericaceous mix).

Perennial

A plant that lives for more than two years. Some plants that are perennial in the wild may be treated as annuals in cultivation.

Petiole

Leaf stalk.

pH

A scale by which the acidity or alkalinity of the soil or compost is measured. The scale runs from 0 to 14, though the extremes are never encountered in horticulture. Although 7 is technically neutral, 6.5 can be regarded as neutral horticulturally, as most plants will grow at this level. Most plants will tolerate a relatively wide pH range, but where particular trees or shrubs need an acid soil this has been indicated in the relevant entry in the first part of the book.

Phototropism

There is a propensity for plants to

grow towards light, which can be a particular problem in a very shady area, where light is usually from one direction.

Plants show varying degrees of phototropism.

Pinching out

This is a form of pruning to encourage the plants to branch out. By removing the growing tip, shoots below are encouraged to branch out to take over. It is a technique usually used for plants that would otherwise tend to become leggy, and always when the plant is relatively young.

Pinna (plural pinnae)

The individual leaflet of a deeply divided leaf, such as most fern fronds.

Planting

It is worth taking time and care over planting as you will only have one chance to get things right.

Dig a hole large enough to spread the root system fully, and mix a bucketful of peat and a handful of bonemeal into the excavated soil (which will be returned later).

Loosen any compacted ground at the bottom of the hole, and if the soil is poor or impoverished, lightly fork in some peat and bonemeal.

Set the tree or shrub in the hole to the same depth as it was before (place a cane across the hole and measure the old soil mark against this if you are in doubt).

If staking is required, drive the stake into the hole *before* planting. *See* Staking.

If planting a container-grown plant, tease some of the roots out if they have started to run around the bottom or edge of the container. If bare-rooted, trickle some soil among the roots then move the plant up and down gently to settle it before adding more soil.

Firm the soil well, and finish off by levelling and then loosening the surface to avoid compaction.

A mulch of peat or pulverised bark will help to conserve moisture.

Plunge

Act of burying a pot up to its rim in the ground. This makes it less liable to dry out and makes watering less critical.

Instead of plunging the pots into the ground, an area can be boarded off and filled with peat or grit for the same purpose.

Pollarding

Process of repeatedly pruning back growth very severely to the main trunk, repeating this every few years.

Pot-bound

A term used to indicate the stage at which the pot has become so full of roots that growth of the plant is suffering. Most plants need repotting once the compost appears full of roots.

Pots

Pots now come in many shapes and sizes—including square—not to mention colours.

Most pots nowadays are plastic, and for many plants these are perfectly suitable. The compost is less likely to dry out so quickly, and they wipe clean easily so unsightly deposits on the inside are unusual.

Be cautious of colours—some will be affected by strong sunlight.

Do not dismiss clay pots. They will make overwatering less likely, and for plants with large or heavy top growth the extra weight might be a useful counterbalance in windy weather.

Potting on

Moving a plant on into a pot of a larger size, usually one or two sizes up. Never pot on into a pot much larger than the present size, otherwise the compost may become either exhausted or 'sour' before the plant has chance to use it.

Pot up

Move a seedling or rooted cutting into a pot for the first time.

Prechill

Tree and shrub seeds are generally much slower and more difficult to germinate than say annuals or most bedding plants. It is often necessary to break their dormancy or to remove germination inhibitors.

Traditionally this was done by sowing them in autumn or early winter and subjecting them to a period of cold outdoors before trying to germinate them, and this has been recommended for many of the plants described in this book.

Nature can be speeded up another way with the aid of a refrigerator.

Place the seeds between *moist* blotting paper, thickly folded kitchen roll, or a piece of flannel, in a container with a lid. Leave this in the refrigerator for two or three weeks, making sure the seeds are kept moist but not waterlogged, then sow in warmth.

Prechilling can be tried for all those plants for which a cold period has been suggested.

Presoaking

Some seeds will germinate more readily if soaked for 12 hours before sowing. Soaking in tepid water has been suggested for several plants in this book. Obviously without special facilities it is difficult to keep the water tepid for that length of time, but no harm will come if you simply change the water as often as possible during that time.

Pricking out

A term used to describe the job of lifting individual seedlings from the box or pot in which they were sown and planting them into individual pots or spacing them out in another box.

Propagation

Act of raising new plants by seed or vegetative means (cuttings, layering, for instance).

Propagator

Cabinet for germinating seeds, raising seedlings, or rooting cuttings. Most are heated to provide the necessary warmth, and enclosed to maintain a very humid atmosphere.

Remember that a propagator must be placed in good light (though certain seeds may germinate better in darkness, the seedlings will need a light position). Avoid direct sunlight, however, as this may increase the temperature too much and cause scorching.

Raceme

An elongated unbranched flower cluster, the individual flowers being stalked.

Repotting

The term repotting is loosely used to mean putting the plant into a new pot, whether the same size or larger. More specifically it means replanting in the same-sized pot after removing some of the old compost and replacing with new. Moving the plant on to a larger pot is more correctly 'potting on'.

Rest/resting period

Many plants have a natural resting period, when they are either dormant or making little or no new growth. This is not necessarily accompanied by a loss of foliage.

Root-ball

The mass of roots and compost when a plant is removed from its pot.

Rootstock

The plant onto which scion material is budded or grafted. The rootstock can affect the vigour and size of the final plant.

Rosette

An arrangement of clustered leaves radiating from a central area.

Runner

A long shoot sent out by some plants, that will root and form new plants where it comes into contact with the soil. This is one of the means by which some plants spread and form ground cover. It is also a very easy way of propagating these particular plants.

Scale

Scale insects look like small yellowish-brown insects that resemble small scales or shells. They are immobile. You can wipe them off with a sponge soaked with an insecticide (but use waterproof gloves). You can also try a systemic insecticide.

Scion

A bud or shoot from the donating plant that is budded or grafted on to a rootstock.

Seeds, plants from

Most trees and shrubs can be raised from seed, though it is often a slow process. It does, however, solve the problem if you do not already have a plant that you can layer or from which you can take cuttings. Some general seedsmen offer a limited range of tree and shrub seeds, but you may have to go to specialist seedsmen for most of them.

Hybrids and varieties of most trees and shrubs are unlikely to come true from seed, and the method is most suitable for the species.

If only a few plants are needed, sow the seeds thinly in a pot of seed compost. If a number are required, then prepare a seed tray as shown.

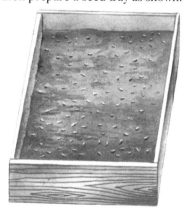

Level the surface of the compost, firming it gently, and space the seeds out well. Cover with about the seed's own depth of compost.

Place the pot in a bowl of water to let moisture soak up from beneath or water the seed tray.

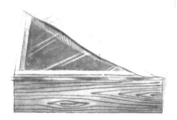

Then cover with a piece of glass, if you do not have a propagator, and a sheet of paper to exclude light.

Once the seedlings have germinated, prick them out into another seed tray or individual pots.

Semi-ripe

Term used to describe a type of shoot used for cuttings. These are the current season's shoots in late summer, when growth has slowed down and the wood started to mature. *See* Cuttings.

Short-day plants

Plants that flower as the nights become longer and the days shorter. They do not normally flower unless it is daylight for less than 12 hours for a period of time.

Shrub

A woody plant without a single tree-like trunk. Some shrubs can be grown as trees, depending on how they are trained.

Species

An individual member of a genus. *See* Genus.

Specimen plant

A tree or shrub (usually) planted to be viewed from all angles, not as part of a large group.

Staking

Trees up to 1.2 m (4 ft) tall when planted do not usually need staking if they are well firmed. Generally it is much better to buy very small plants than large ones. If you do plant a tall sapling, a stake will be necessary for the first 18 months—maybe for another season if the plant is over 2.4 m (8 ft) tall when you plant it.

A short stake is usually perfectly adequate—there is no advantage in having a stake that extends above the main stem of the tree, and one that reaches half way up the clear stem should be long enough.

Drive the stake into the hole before planting the tree, then it will not damage the roots when you drive it in.

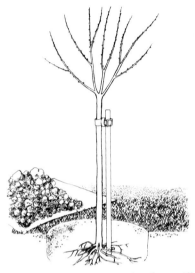

Always use a tree-tie that will hold the tree away from the stake.

Stamen

Male reproductive organ, which forms and carries the pollen.

Stool

A tree or shrub as grown as a cluster of young stems arising from ground level, achieved by cutting the shoots back close to ground level annually.

The term is also used to describe the crowns of herbaceous plants that are lifted annually for propagation, such as chrysanthemums.

Stop/stopping

To 'stop' a plant is to pinch out the growing tip to encourage lateral growths to develop. *See* Pinching out.

Stratify

To break dormancy of seeds by exposing them to low temperatures before sowing. Fleshy fruits are crushed first, then placed in layers of sand or peat in pots or boxes and stood outdoors (protected from vermin) during the winter. The seeds are cleaned and then sown in warmth. *See also* Prechilling.

Strike

To root a cutting.

Systemic

A term used to describe insecticides and fungicides that can be taken up by the plant and translocated (moved about) within the plant. The chemical can be taken up by the roots and will find its way to the leaves.

Tendril

A thread-like modified leaf or stem that can twine round a support to enable the plant to cling and climb.

Tree

A woody plant with a distinct trunk or main stem, though some trees have several trunks. Some trees can be grown as shrubs, depending on the initial training.

Tree-tie

A tie that holds the tree securely close to a stake, but keeps it out of direct contact with the stake to avoid friction damage. Use tree-ties that are adjustable, so that they can be slackened as the stem grows.

Several types are available, usually made of plastic. The kind with a buckle is easy to adjust.

Trifoliate

A leaf divided into three leaflets.

Umbel

A cluster of flowers in which all the individual stalks arise from a common point at the top of the main flower stem.

A compound umbel is an umbel of umbels.

Vegetative propagation

Any method of increasing a plant other than by seed.

Weeping tree

A tree of pendulous habit, with branches that cascade towards the ground.

Whorl

An arrangement of leaves or flowers arising from one point, arranged rather like the spokes of a wheel.

Whip-and-tongue graft

A type of graft. *See* Grafting.

Witches' brooms

Twiggy outgrowths sometimes found on trees, caused by irritation of the tissue. They are not unusual on birches, and are likely to be caused by insects or fungus infections. Cut out the infected part, taking the branch back to 15 cm (6 in) below the broom. Coat the cut area with a wound paint.

Common Name Index

The most important Latin synonyms are also included

Picture credits
H. Allen: 118(tr)
Amateur Gardening: endpapers
D. Arminson: 45(t)
P. Ayers: 62(t), 85
G. Beckett: 42, 73(b), 97
T. Birks: 76(t)
Pat Brindley: 37
R.J. Corbin: 62(b), 75, 78(b), 106(r), 111(r), 114(l,c,tr), 118(l)
J. Downward: 44(b)
Alan Duns: 9, 24, 25(t), 31, 33, 38(b), 40(t), 79, 91, 94(t)
Valerie Finnis: 27(b), 30(t), 60
P. Genereux: 45
Peter Hunt: 74(t), 105(tr), 106(l)

George Hyde: 35(r), 36, 98(t), 118(br)
Peter McHoy: 7, 8, 10(t,b), 12(t,b), 14(t), 16(l,r), 19(t), 20(t,b), 21(t), 22(t), 23(b), 28(b), 29, 38(t), 39, 43, 44(t), 49, 50(b), 61(b), 65, 77, 84(b), 86(t), 89, 90(b), 100(t)
M. Newton: 53(t), 90(t)
M. Nimmo: 35(l)
S.J. Orme: 111(l)
R. Perry: 76(b)
N. Procter: 117(b)
Rosenwald: 56(t)
Ianthe Ruthven: 54
Miki Slingsby: 15, 19(b), 27(t), 32(t), 58(t), 78(t)

Harry Smith Horticultural Photographic Collection: 2, 11, 13, 14(b), 17(t,b), 18, 21(b), 23(t), 25(t) 26(b), 28(t), 32(b), 34, 41(t,b), 46(t,b), 47, 48(b), 50(t), 51, 52, 53(b), 55(t,b), 57(l,r), 58(b), 59, 63, 64(b), 66(t,b), 67, 69(b), 70, 71(t,b), 73(t), 80, 81, 82(b), 86(b), 87, 88(t,b), 93, 94(b), 95, 96(l,r), 99, 100(b), 101, 104(bl), 110, 118(l)
Michael Warren: 1
Colin Watmough: 6, 22(b), 25(b), 30(b), 40(b), 56(b), 61, 64(t), 69(t), 72, 74(b), 82(t), 92(t,b), 98(b)